Cambridge Elements ≡

Elements in Politics and Society in East Asia
edited by
Erin Aeran Chung
The Johns Hopkins University
Mary Alice Haddad
Wesleyan University
Benjamin L. Read
University of California, Santa Cruz

ENVIRONMENTAL POLITICS IN EAST ASIA

Mary Alice Haddad
Wesleyan University

CAMBRIDGE
UNIVERSITY PRESS

Shaftesbury Road, Cambridge CB2 8EA, United Kingdom

One Liberty Plaza, 20th Floor, New York, NY 10006, USA

477 Williamstown Road, Port Melbourne, VIC 3207, Australia

314–321, 3rd Floor, Plot 3, Splendor Forum, Jasola District Centre,
New Delhi – 110025, India

103 Penang Road, #05–06/07, Visioncrest Commercial, Singapore 238467

Cambridge University Press is part of Cambridge University Press & Assessment,
a department of the University of Cambridge.

We share the University's mission to contribute to society through the pursuit of
education, learning and research at the highest international levels of excellence.

www.cambridge.org
Information on this title: www.cambridge.org/9781108965774

DOI: 10.1017/9781108966085

First published 2023

A catalogue record for this publication is available from the British Library

ISBN 978-1-009-45436-0 Hardback
ISBN 978-1-108-96577-4 Paperback
ISSN 2632-7368 (online)
ISSN 2632-735X (print)

Environmental Politics in East Asia

Elements in Politics and Society in East Asia

DOI: 10.1017/9781108966085
First published online: September 2023

Mary Alice Haddad
Wesleyan University

Author for correspondence: Mary Alice Haddad, mahaddad@wesleyan.edu

Abstract: Once viewed as an environmental hazard to the planet, East Asia is now at the forefront of pro-environmental policymaking. The region's progress has been both remarkable and surprising given the pro-business orientation of its political systems and their ideological diversity. Through a focus on three environmental policy areas exhibiting different levels of success, this Element shows how governments in China, Japan, South Korea, and Taiwan have been able to craft pro-environmental policy by working in collaboration with business and societal interests. The evolution of the region's eco-developmental states has resulted in exceptional progress in the areas of green technology and green finance, mixed outcomes in pollution management, and negligible improvement in addressing environmental justice. As the planet seeks guidance in addressing our collective climate crisis, East Asia offers both hope and caution for how we can craft pro-environmental policies in diverse political contexts.

Keywords: environment, East Asia, climate, business, policy

ISBNs: 9781009454360 (HB), 9781108965774 (PB), 9781108966085 (OC)
ISSNs: 2632-7368 (online), 2632-735X (print)

Contents

1 Introduction: East Asia's Environmental Politics

Once viewed as an environmental hazard to the planet, East Asia is now at the forefront of pro-environmental policymaking. The region's progress has been both remarkable and surprising given the pro-business orientation of its political systems and their ideological diversity. Nationally viable green parties and a strong regional governance system, which were largely responsible for the spread of pro-environmental policies in Europe, are absent in East Asia. Similarly missing is a rich and powerful nonprofit/public interest sector, which advocated for environmental policy change in the United States. In contrast, East Asia is known for its influential business sector, which has traditionally been opposed to pro-environmental policy.

Thus, the core questions of this Element are: How and why did East Asia's ideologically diverse pro-business governments transform their policy orientations from a highly polluting growth-at-any-cost to ones that are much more eco-friendly? This Element argues that East Asia's path toward pro-environmental policy has been led by national government bureaucracies in close collaboration with business, both of which are continually shaped by citizen interests and expectations.

Based on more than ten years of research on environmental politics in East Asia, this Element uses a combination of quantitative and qualitative methods to focus on three policy areas – green business, pollution, and environmental justice – in China, Japan, South Korea, and Taiwan. It will argue that public pressure to address pollution that was poisoning communities and destroying their livelihoods was the initial spark that started pro-environmental policymaking in the region. However, it was the business sector's recognition that pro-environmental policy can also be pro-business that allowed piecemeal anti-pollution regulations to be transformed into comprehensive pro-environmental policies.

As the business sector discovered the commercial benefits of cleaner production processes and began to price the risks of climate change into its long-term growth projections, it began to become more vocal about the benefits of pro-environmental policies. Through the combined efforts of connected stakeholders across academia, local and national governments, citizen groups, and business, the political tide favoring the environment shifted in East Asia (Haddad 2021). The specific political configurations were slightly different in each place, but the collective results were very similar: growth-at-any-cost developmental states modified their policies in ways that incorporated environmental considerations into their policymaking and became eco-developmental states (Esarey et al. 2020). Now, East Asia's eco-developmental states are prioritizing environmental considerations across a wide range of domestic policymaking and are actively working to spread the eco-developmental model abroad.

My focus on these three policy areas is intended to highlight the ways that East Asia's pro-business approach works better in some areas of environmental policy than in others. In particular, when pro-environmental policies can generate profits with relatively little cost, such as green technology, green finance, and clean energy, policymaking in the region has moved the furthest and been the most progressive. When environmental benefits require businesses to incur costs, even if they also reap benefits, such as cleaning up production processes and mitigating pollution, policymaking has been more difficult and the outcomes more mixed. Finally, in environmental issue areas that generate very little profit but require significant investment and political will, such as addressing environmental justice, East Asia's pro-business model of environmental policymaking has often failed.

Therefore, while this Element offers a positive example of East Asia's pro-business path toward a more sustainable future, the example comes with a caveat. All of East Asia's current leaders have made public commitments about transforming their economic systems in ways that prioritize environmental protection and social progress. In his statement at the September 2020 United Nations (UN) Summit on Biodiversity, China's President Xi underscored the importance of "stay[ing] the course for green, inclusive and sustainable development."[1] Japan's Prime Minister Kishida has reiterated his commitment to reshaping Japan's economy, promoting what he called a "new form of capitalism" in which "social challenges become the engine of growth."[2] In his 2022 keynote address to the UN General Assembly, South Korea's President Yoon Suk Yeol underscored that "broadening support for the socially disadvantaged groups lays the groundwork for sustainable prosperity."[3] Finally, in the inaugural address of her second term, President Tsai committed to making "Taiwan a center of green energy" while enhancing a social safety net that would create a "new era of shared prosperity."[4]

---

[1] Ministry of Foreign Affairs website: http://sl.china-embassy.gov.cn/eng/xwdt/202010/t20201003_5848841.htm.

[2] "PM Kishida's Speech on the New Form of Capitalism and Why Japan Is a 'Buy'," *Kizuna* (JapanGov), June 23, 2022, www.japan.go.jp/kizuna/2022/06/why_japan_is_a_buy.html.

[3] "President Yoon Suk Yeol, Keynote Speech at the 77th UN General Assembly," Ministry of Foreign Affairs News (Republic of Korea), September 21, 2022, www.mofa.go.kr/eng/brd/m_5674/view.do?seq=320741.

[4] "Inaugural Address of ROC 15th-Term President Tsai Ing-wen," Office of the President, https://english.president.gov.tw/News/6004#:~:text=So%20over%20the%20next%20four,our%20country%20into%20the%20future.&text=I%20know%20that%20the%20Taiwanese,our%20indus

In sum, this Element argues that the East Asian experience demonstrates that very different types of governments have been able to shift their policies in pro-environmental ways even as they have retained their pro-growth and pro-business orientation. Their experience may inspire hope for countries struggling to achieve economic development without killing the planet. They also provide a cautionary tale about the costs that such a pro-business path might entail.

1.1 Why Study Environmental Politics in East Asia?

There are two compelling reasons to examine East Asia when studying environmental politics. The first is that the region is so large in terms of population, economy, and carbon emissions that efforts to combat global climate change cannot succeed without it. East Asia contains 22 percent of the total world population,[5] 30 percent of global gross domestic product (GDP),[6] and 34 percent of global carbon emissions.[7] Therefore, if East Asia can successfully modify its economy to one that is sustainable, the world has a chance of averting the impending climate disaster. If East Asia fails, the world will fail.

The second reason why the region is useful to study is because it offers an unusual combination of commonalities and variation, making it possible to examine how a number of different factors affecting environmental politics actually influence outcomes. China, Japan, South Korea, and Taiwan all share a somewhat related cultural background rooted in Confucianism. All four places experienced a period of "high-growth" rapid industrial development that relied heavily on export-oriented industrial structures supported by pro-business governments. All four legal systems significantly restrict political advocacy, with the predictable result that none of them have nationally viable green parties or large advocacy sectors. Residents in all four places have experienced significant, intense levels of pollution. All four have seen citizens and their organizations demand that their governments address pollution problems, and governments and businesses in all four places have responded with stronger environmental policies. These commonalities mean that some of the most widely accepted factors that the literature expects to influence environmental politics can be considered as roughly equivalent across the region.

---

trial%20and%20economic%20development.: *The China Post* website, https://chinapost
.nownews.com/20200520-1266619.

[5] Eastern Asia Worldometer data, www.worldometers.info/world-population/eastern-asia-population/#:~:text=The%20current%20population%20of%20Eastern,among%20subregions%20ranked%20by%20Population.

[6] World Bank data 2019, https://data.worldbank.org/indicator/NY.GDP.MKTP.CD?locations=Z4-1W.

[7] Calculated from International Energy Agency (IEA) 2018 data, www.iea.org/data-and-statistics?country=WORLD&fuel=CO2%20emissions&indicator=TotCO2.

While they share many commonalities, China, Japan, South Korea, and Taiwan also vary in some very important ways. The most obvious difference is size. China is the most populous country (1.4 billion people) and has the second largest GDP ($18 trillion) in the world. In contrast, Taiwan is comparatively tiny, with only 24 million people and a GDP of $791 billion.[8] Japan (population of 126 million, GDP of $5.2 trillion) and South Korea (population of 52 million, GDP of $2.4 trillion) sit between those two extremes.[9]

Another critical way the four differ is their political regimes. Japan is the oldest democracy in the region. Its 1947 democratic constitution guaranteed its citizens equality under the law, due process, and freedom of expression, assembly, and religion. South Korea and Taiwan are newer democracies. They both experienced lengthy and sometimes brutal occupations by Japan (South Korea from 1910 to 1945 and Taiwan from 1895 to 1945). Postwar, they suffered destructive civil wars that split their prewar territories in two, with one part controlled by communists and the other by nationalists. Both South Korea and Taiwan democratized in largely peaceful revolutions in the late 1980s, developed robust two-party political systems, and have experienced several peaceful alternations in power.

Finally, mainland China is not democratic. Its constitution uses the phrase "dictatorship of the proletariat" to describe its political system. While China's constitution grants its citizens equality under the law, freedom of the press, and freedom of assembly, religion, and so on, those individual rights are subordinate to the rights of the state (Article 51). Similarly, although local elections are frequently competitive, candidates have been preselected to limit voter choices to those acceptable to the Chinese Communist Party (CCP), and citizens do not have the right to elect their national leaders directly (Manion 2000). The traditional press is highly restricted, and while significant freedom is allowed on the Internet, it too is frequently censored (King, Pan, and Roberts 2013; Xiao 2019). Finally, after a period of political opening in the 2000s, in which civil society and the private business sector grew and flourished, over the last decade the government has greatly increased its authority and oversight over social, political, and economic life, constraining the capacity of non-CCP actors to influence policy (Naughton 2017; Snape 2021).

---

[8] China World Bank GDP data, https://data.worldbank.org/indicator/NY.GDP.MKTP.CD? locations=CN; China World Bank population data, https://data.worldbank.org/indicator/SP .POP.TOTL?locations=CN. Taiwan GDP data: https://www.imf.org/external/datamapper/pro file/TWN; Taiwan population data: https://statisticstimes.com/demographics/country/taiwan-population.php.

[9] "Economy of Japan," Wikipedia, https://en.wikipedia.org/wiki/Economy_of_Japan; "Economy of South Korea," Wikipedia, https://en.wikipedia.org/wiki/Economy_of_South_Korea.

Therefore, studying East Asia allows us to control for many of the factors that the environmental politics literature has posited should influence environmental politics, such as green parties (O'Neill 1997), nongovernmental organization (NGO) strength (Bosso 2005), and economic growth and trade patterns (Dasgupta et al. 2002; Prakash and Potoski 2007), while allowing us to explore the ways that regime type and size might influence the evolution of environmental politics in the region.

2 Eco-developmental States

This Element argues that East Asia's environmental politics is governed by its eco-developmental states. The concept of an eco-developmental state was introduced in Esarey and colleagues' (2020), *Greening East Asia: The Rise of the Eco-developmental State*. The term is meant to capture a governance structure in which states are active in developing international competitive advantage, collaborate closely with business in making policy, enjoy the broad support of their societies, and have prioritized the environment as a top policy goal. This section will describe how the concept evolved as well as its key features.

2.1 Evolution of a Concept

The concept of the developmental state emerged with the publication of Chalmers Johnson's (1982) now classic book *MITI and the Japanese Miracle*, which argued that the key to Japan's extraordinary economic growth lay in its governance structure. Johnson argued that Japan's developmental state followed neither a market-rational model, as exemplified by the United States, in which prices and production levels were determined by markets, nor a planned-economy model, as exemplified by the Soviet Union, in which businesses were state-owned and prices and production levels were generally set by the central government (Johnson 1982). Instead, in Japan's developmental state, business remained private, and prices and production levels were determined by market forces, as they were in the United States, but also an elite government bureaucracy was active in identifying strategic industries and guiding business toward expanding those industries (Johnson 1982).

As Japan's industrial policy became better understood and its model came to be adopted by other countries in the region and elsewhere, the developmental state model came to be called a number of different names, including a "coordinated market economy" (Hall and Soskice 2001), "strategic capitalism" (Calder 1995), and "welfare capitalism" (Dore 2000), to name a few. Whatever it is named, from a policymaking standpoint, the key features of the model are institutional structures that allowed the bureaucrats and counterparts outside the government (usually business, sometimes NGOs) to coordinate with one another when making policy. As Stephan Haggard explains in his Element *Developmental States*, "'deliberation councils' linking business and government played an important role in resolving credibility problems associated with authoritarian rule and building trust between the public and private sectors" (Haggard 2018: 54).

From an economic standpoint, the developmental state was exceptionally successful, generating very high growth rates over several decades, which lifted hundreds of millions of people out of poverty, raised life expectancies, expanded access to health and education, and dramatically improved the living standards of nearly everyone in the entire region. However, rapid growth based on industrial output had a significant environmental cost. Within about twenty years of launching their rapid industrialization, developmental states would all face significant social, political, and social crises as a direct result of industrial pollution.

As discussed in more detail in Section 3, growth-at-any-cost developmental states were no longer viable ecologically, politically, or economically. Ecologically, soil, water, and air became toxic, poisoning both workers and customers. Politically, citizens demanded that their governments prioritize residents' health and welfare along with corporate profits. Economically, the agricultural industry was decimated by pollution and global manufacturers faced markets in the United States and Europe that were restricting the imports of hazardous products and goods manufactured using polluting processes. As a result, developmental states were forced to begin including environmental considerations into their industrial policy and economic growth plans (Esarey et al. 2020).[10]

Scholars discussed the early accommodations of anti-pollution activists' concerns as regular policy adjustments were made in responses to domestic political pressure. While some reforms were significant – such as the creation of new agencies (e.g., in 1971, Japan created an Environment Agency and Taiwan formed its Environmental Protection Administration; South Korea's Environment Administration was formed in 1980; and China's State Environmental Protection Administration formed in 1998) and laws (e.g., South Korea's Environmental Pollution Prevention Act of 1963, Japan's 1967 Basic Law for Environmental Pollution Control, Taiwan's Water Pollution Control Act of 1974, and China's Air Pollution Control Act of 1987) – they

---

[10] Recently, Schaede and Shimizu (2022) have attributed the same set of observed shifts in the developmental states' policymaking process and policies to technological and demographic pressures/opportunities.

were not viewed as fundamentally changing the nature of East Asia's developmental states. Scholars studied domestic environmental protests and movements and examined them as analogous to interest groups and political movements elsewhere that pressured governments to include environmental priorities in legislation and policy implementation (Broadbent 1998; Ku 1996; Mertha 2008; Reardon-Anderson 1997).

In the cases of Japan and China, the ruling parties (the Liberal Democratic Party [LDP] in Japan and the CCP in China) made policy adjustments to accommodate some public concerns about pollution, but they did not really include opposition political actors at the heart of policymaking. In both countries, environmental advocates made policy gains by accessing the government through administrative and legal processes. Laws and regulations at both the national and the local level were generally developed by policy advisory committees that included pro-environment academics serving as technical experts who advocated for more stringent pollution controls (Jing 2003; Schreurs 2002). Both countries also used (in the case of Japan) and created (in the case of China) legal channels that enabled pollution victims to seek redress through the state-controlled venue of the courts (Upham 1987; Wang 2013). For the most part, environmental advocates from the 1980s onwards in both countries found ways to work within the system, making allies with policymakers in the government. They treated environmental protection as a policy issue with technical and administrative solutions, and they avoided making it a partisan issue (Hildebrandt and Turner 2009; Imura and Schreurs 2005).

The story in South Korea and Taiwan was quite different. In those two countries, anti-pollution movements merged with pro-democracy movements in the 1980s, seeking an end to the nationalist, pro-military governments that dominated politics. Relatively peaceful democratic transitions followed – South Korea adopted a new constitution in 1987 and Taiwan ended martial law the same year. In the context of the democratic transition, environmental activists joined liberal parties to advocate for change through the political process. In the years that followed, both places saw the emergence of a two-party system where liberal party-led governments tended to include pro-environmental activists in key political posts and conservative party governments tended to exclude them. Thus, even after democratization, while South Korea and Taiwan's conservative governments made some policy accommodations to address the public's pollution concerns, the basic growth-first policymaking structures of their developmental states did not change very much (Haddad 2015; Ho 2010; Ku 2011).

The evolution from developmental state to eco-developmental state in East Asia happened over the course of two decades and had two somewhat distinct phases – the first was led by the NGO sector and the second was led by business. The early years of the 2000s saw a slow but dramatic shift across the region (and the world) in the ways that environmental concerns were incorporated into policy and pro-environmental actors were included in decision-making. Environmental NGO professionals were more regularly incorporated into decision-making at the highest levels in all countries. International relations scholars attributed the change to the rise of global civil society, the professionalization of NGOs, and the dramatic improvement in communications technologies. Together, these factors significantly empowered networks of pro-environmental actors in academia, civil society, and governments around the world to collaborate in pushing pro-environmental policy change in their respective countries and localities (Keck and Sikkink 1998; Rodrigues 2003).

These broader global trends were both facilitated by East Asian actors and felt within East Asian governments. The 1997 summit of the Conference of the Parties of the UN Framework Convention on Climate Change, which resulted in the Kyoto Protocol, served as a focal point for global environmental organizing as well as an opportunity for the Japanese government to begin giving NGO professionals a seat at the policymaking table (Reimann 2003) and highlighting sustainable development as a foreign policy priority (Schreurs 2005). China was increasing its role in international organizations (it joined the World Trade Organization [WTO] in 2001) and found global environmental NGO professionals helpful partners as it sought to navigate international politics and develop domestic regulations that would attract foreign aid and investors (Economy 2004; Hildebrandt and Turner 2009; Yang 2005).

The second evolutionary phase in East Asia's eco-developmental states can be marked by the 2008–9 global financial crisis and, more importantly, the recovery that came afterwards. Prior to the crisis, global NGOs had increased their size, capacity, and political clout and were demanding more transparency from multinational corporations, calling for consumer boycotts, and rewarding firms with greener records. As a result, businesses around the world, including East Asia's multinational corporations, were discovering that offering "green" products could be more than just good public relations; they could enhance a company's bottom line. In 1996, the International Standards Organization began publishing the ISO 14000 series of standards that enabled companies anywhere in the world to demonstrate that their products and/or processes met internationally recognized standards. Gradually, companies found that voluntarily participating in environmental programs could increase sales and investors (Prakash and Potoski 2007).

When the 2008–9 global economic crisis hit, governments across the region (and the world) rolled out gigantic financial stimulus packages, and East Asia's governments all turned to a variety of "green growth" strategies to revive their sagging economies in the wake of the crisis. South Korea's plan was the first and most ambitious (Zelenovskaya 2012), but all the countries in the region made significant investments in renewable energy, electric vehicles, intensive and organic agriculture, and similar industries to spur competitive development in important future industries while reviving the economic prospects of sagging rural and industrial areas (Barbier 2010; Fardoust, Lin, and Luo 2012; Ladislaw and Goldberger 2010). These trends were only enhanced toward the end of the decade with the dramatic expansion of green finance (more on green finance in Section 4).

By the end of the 2010s, East Asia's states had undergone sufficient political transformation that scholars began to question whether it was still appropriate to call them "developmental states" (Pirie 2018; Williams 2014). Others, noting the enhanced attention to environmental priorities alongside a continuation of many developmental state decision-making structures and methods, developed terms like "developmental environmentalism" (Kim and Thurbon 2015) and "eco-developmental state" (Esarey et al. 2020) to capture the use of developmental state industrial policy methods to target pro-environmental policy priorities.

2.2 Defining Features

This Element further develops the eco-developmental state concept presented in *Greening East Asia* (Esarey et al. 2020). In his overview of important developmental state literature, Ziya Öniş suggests that there are three key features of East Asia's developmental states that made them economically successful: (1) "single-minded adherence to growth and competitiveness," (2) high level of bureaucratic autonomy and capacity, and (3) high level of public–private cooperation (Öniş 1991: 120). These features enabled developmental states to engage in a variety of strategic state interventions in the markets, generally by supporting key industries that would enhance national global competitive advantage (Amsden 1992; Vogel 1996; Wade 1990).

In their evolution from "pure" developmental states to eco-developmental states, East Asian governments softened their "single-minded" pursuit of rapid economic growth to one that includes environmental concerns alongside more moderate economic growth goals. As discussed in Section 2.1 and as will be elaborated more in Section 3, this shift came about not because of a moral awakening; rather, it emerged directly from the political pressure exerted by pollution victims and the economic reality that toxic products and processes hurt the long-term profitability of firms. By the end of the 2010s, East Asian governments had adjusted their earlier developmental state structures to accommodate the new reality that (1) they were leading rather than following global economic development, (2) markets had become complex and supply chains were global, and (3) long-term profits required industrial policies that were sensitive to the environmental, social, and political concerns of local populations.

Rather than declare East Asian developmental states "dead," this Element argues that they evolved into eco-developmental states. These newer versions retain several key characteristics of developmental states but have added some new goals and new players to their policymaking systems. These modifications are significant enough that they constitute a new evolutionary form of the developmental state.

Perhaps the primary defining feature of developmental states is their focus on "competitive" rather than "comparative" advantage when building their industrial policy. Popularized by Michael Porter in *The Competitive Advantage of Nations*, the idea of national competitive advantage is that the factors that help countries succeed the most in international trade are created by governments and businesses and not simply inherited (Porter 1990). Countries that invest heavily in human capital and foster clusters of competitive firms in industries with global markets can grow their national economies through trade irrespective of what assets they may have at the start. Countries building international competitive advantage will take full advantage of their "natural" endowments (e.g., natural resources, cheap labor, access to shipping lanes, etc.), but these will not be the most important components of their industrial growth strategy.

Initially, developmental states were countries that sought to "catch up" with advanced industrial economies. Thus, they were in the advantageous position of being able to learn what the most important growth-generating industries were from the experience of others and target those industries for government support. Furthermore, crafting competitive advantage was not seen as a task for only the trade and finance ministries but was a whole-government effort. Education, labor, welfare, health, transportation policies and more were all coordinated to support the government's industrial development goals (Esping-Andersen 1990; Hall and Soskice 2001).

A second defining feature of developmental states is the close collaboration between government and business when developing policies and designing policy systems to support them. Peter Evans used the term "embedded autonomy" to describe the relationship that developmental states' bureaucracies had with their societies, especially the business sectors. These bureaucracies had high administrative capacity and the power to make decisions that were not unduly influenced by special interests while at the same time being highly connected to important economic actors (e.g., business and labor), enabling them to make policies that could support rapid industrialization and high economic growth without causing civic unrest or being captured by corrupt special interests (Evans 1995). At a practical level, developmental states generally made policy by relying on advisory committees that included capable bureaucrats, technical experts (often from academia), and someone who could speak about the concerns of labor and the public (Schwartz 1998; Vogel 1996).

A final feature of developmental states is broad social support for a government's goals of rapid economic development. Rapid industrial development causes significant social changes – millions of people move from the countryside to the cities, labor shifts from working in the fields to working in factories, families are smaller, children spend most of their days in school instead of at home or outside, and so on. Individuals and families are required to make big, often unpleasant, changes in their lives to make rapid growth happen in the country. Developmental states craft a social bargain with their citizens that can be simplified as work hard now and have a better life for yourself and, especially, your children later. Since developmental states were "catching up," these citizens could look to the already advanced industrialized states for a good idea of what they might be able to get in return for their sacrifices: consumer products like cars and washing machines, leisure time they could spend on hobbies and vacations, longer and healthier lives. Citizens in developmental states accepted this bargain, supported rapid industrial development, and many of them and nearly all of their children received the increased standard of living that their governments had promised.

Eco-developmental states closely resemble developmental states, especially in terms of these three key features. All three foundational components have been modified significantly, which is why I call contemporary East Asian states eco-developmental states rather than developmental states. Eco-developmental states are still focused on developing their competitive advantage. They continue to invest heavily in human capital and direct government resources to support clusters of industries to help them compete in global markets. These efforts continue to be comprehensive ones that involve coordination across government ministries and between central and local governments.

While supporting and building competitive advantage remains a top priority for eco-developmental states, they no longer have the myopic growth-at-any cost view of their predecessors. Eco-developmental states recognize that rapidly running up a road already traveled by many predecessors is not the same as charting a new road. Eco-developmental states are now leading rather than following global economic innovation. Not only do they move forward more slowly, but they must also take more time to figure out where they want to go and devote more energy planning how to get there.

Whereas developmental states focused almost exclusively on industrial development, eco-developmental states seek to develop cutting-edge industrial capacity as well as robust service and information sectors. Eco-developmental states do not accept that environmental protections necessarily hurt profits. Instead, they see environmental products and services as new markets with large global growth potentials and recognize that polluting products and processes cost everyone over the long term. A key feature of an eco-developmental state is that environmental protection and climate change are top policy priorities and environmental industries and services are among the top recipients of governmental support.

Eco-developmental states continue to work closely with business as they develop and implement their growth strategies. While big businesses retain privileged access to policymaking, policy advisory committees have members connected to diverse networks of stakeholders. Eco-developmental state technocrats seek advice not only from old friends in global manufacturing but also from start-up entrepreneurs, nonprofit organization (NPO) professionals, and academics who bring with them both technical expertise and experience volunteering for community-based groups and NGOs (Haddad 2021). A key feature of eco-developmental states is the regular inclusion of people connected to pro-environmental networks in policymaking processes.

Finally, like their predecessors, eco-developmental states have crafted social contracts that have earned them the support of their populations, although those bargains look different than they did in earlier decades. For the most part, developmental states made good on their promises of giving their citizens a better life in exchange for hard work. To take just one crude but telling measure, in 1950 life expectancies in East Asia were very low: South Korea's was just twenty-one years, China's was forty-four years, Taiwan's was fifty-six years, and Japan's was fifty-nine years. Children born in 1950 in the United States could expect to live many decades longer; their life expectancy was sixty-eight years. By 2020, children born in East Asia could expect to have among the longest lives in the world, and most would live longer than those in the United States. Life expectancy for children in 2020 was seventy-eight years in China, eighty-one years in Taiwan, eighty-four years in South Korea, eighty-five years in Japan, but only seventy-seven years in the United States.[11]

As will be discussed in greater detail in Section 3, these gains came at a very steep price, requiring renegotiation of the original social contract. Imagine you

---

[11] Data from Our World in Data, https://ourworldindata.org/search?q=life+expectancy.

are in a group of ten cold, hungry people trying to figure out what to do. It will be relatively easy to get all of them to agree that getting warm and fed should be top priorities. However, once you are all warm and fed, it will be much more difficult to gain agreement on what to do next – Go for a walk? Read a book? Start a new business? Have a dinner party? Indeed, whatever you collectively decide as the next activity will probably require the flexibility for different groups of people to do different things.

Similarly, eco-developmental states have crafted social contracts that continue extensive government support for globally competitive industries, but the industries and services gaining governmental support are significantly more varied – by size, type, and geography – than the earlier ones. Additionally, many firms supported by the government will focus on environmental goods and services, and traditional industrial firms will have modified their production processes and products to mitigate negative environmental impacts. A third key feature of eco-developmental states is a social contract that includes broad support for the government in exchange for support of a high standard of living and diverse lifestyle choices while addressing environmental concerns and devoting significant governmental resources to environmental goals.

2.3 Conclusion

East Asia offers a fascinating and important story about how poor, undemocratic countries can develop economically and then expand policy priorities to include environmental considerations. The East Asian experience demonstrates that it is possible for pro-business governments to turn away from growth-at-any-cost development toward developing policymaking structures that include pro-environmental voices in policymaking and target green industry and green services for significant governmental support.

Therefore, the root of the transformation from developmental states to eco-developmental states in East Asia lies with its people, who used their powers as citizens and consumers to raise awareness of the harms of environmental pollution and the benefits of pro-environmental policy and action. The public's shifting pro-environmental interests were then supported by politicians and bureaucrats who crafted state policy to require and encourage better environmental behavior on the part of citizens, companies, and governments. Many companies also supported this shift as they discovered that greener products and processes could improve their brand image as well as their sales and profits.

The precise mechanics of how citizens and corporations shifted governmental policymaking is described in more detail in Section 3. The legacies of pro-growth developmental states mean that each of the countries has a similar configuration of governmental and business entities that influence environmental policy. The real differences among the East Asian countries can be found in the channels through which citizens and civil society actors make their concerns and ideas known to policymakers. Although citizen concerns and political pressure passed through different channels in the different countries, they are all pointing in the same direction, demanding greater attention to human health and greater care for the natural environment.

3 Environmental Politics in East Asia: A Brief History

The environment first became a political issue in East Asia when people living in rural communities found their lives and livelihoods threatened by industrial pollution. Local residents would then organize to try to stop harmful pollution and demand redress from companies and the government. Initial governmental and corporate responses were predictably resistant to change and were often violent in their attempts to coerce communities into accepting the costs of pollution in exchange for a variety of economic and other benefits. When residents refused to be bought off, government and corporate actors engaged in a wide spectrum of responses, ranging from violent suppression, coercion, co-optation, compromise, and even innovation (Economy 2004; Harris and Lang 2015).

While initial movements in the region took the form of classic NIMBY (not in my back yard) struggles, over time they became more sophisticated. Activists found ways to work with local and national governmental officials as well as corporations to gain better environmental outcomes. As they began to work with local residents and governments, corporations found that better environmental practices led to a healthier, happier, and more productive workforce, which in turn led to better products. Eliminating waste often saved money, and cleaner, more energy-efficient products were more popular with consumers, so these pro-environmental changes by companies revealed that many pro-environmental business practices could also lead to higher profits. As the global environmental movement spread, companies and governments both recognized the risks of climate change and could price them into their economic models, further bolstering support for pro-environmental policymaking. Today, East Asian governments are at the forefront of the global effort to build a sustainable future for our planet.

3.1 Japan

Asia's environmental movements began in Japan, which was the first country in the region to industrialize. When the country opened up to international trade in the late nineteenth century, industry dramatically expanded. Its first set of environmental protests erupted in the late 1890s and early 1900s in villages hosting copper mines. Sulfur gas emitted from the smokestacks clouded the air, and heavy metals and acid released into the wastewater poisoned the crops and fish in the surrounding areas, causing serious health problems for residents and harming the livelihoods of nearby farmers and fishermen. Initially, the companies and governments tried to deny responsibility and repress protests. However, company engineers soon discovered that longer smokestacks dispersed the pollution sufficiently to address the local contamination problems. Once the localized pollution was reduced and the families paid off, the protests faded (McKean 1981; Watanabe 2013).

The pattern was repeated fifty years later when Japan's postwar recovery policies favored economic growth over environmental protection, resulting in widespread, highly toxic industrial pollution across the archipelago (Walker 2011). This time, however, Japan was a democracy, and Japan's civil society was connected to the world. Joining counterparts in the United States and Western Europe, Japanese pollution victims took their complaints to the public and their corporations to court. To the surprise of many, they won (McKean 1981; Upham 1976).

The government reacted relatively quickly to the widespread and growing concerns about the environmental costs of its growth-first economic policies. Unlike its prewar predecessors, the conservative Liberal Democratic Party (LDP) was sensitive to electoral pressure from opposition parties and quickly passed a sweeping array of environmental legislation in 1970 in what has come to be known as the Pollution Diet. Companies complied, and from that time forward, corporations worked to stay ahead of pollution issues by proactively initiating voluntary commitments to prevent intrusive government regulations and avoid negative branding.[12]

While the regulations and institutions put in place in the 1970s cleaned up Japan's air and water, they did not address the problem of greenhouse gas emissions and climate change, which only began to be recognized as a problem with the first Earth Summit in Rio de Janeiro in 1992. A few years later, in 1997, two very significant things happened, which together caused a large shift in Japan's approach to environmental politics.

First, the third Conference of the Parties (COP 3) was held in Kyoto, and participants developed the Kyoto Protocol, a landmark international treaty limiting greenhouse gas emissions (as of this writing, 192 countries are party to the treaty).[13] The global environmental advocacy community descended on Japan for the duration of the conference, galvanizing local groups (Reimann 1999).

---

[12] Interviews with senior managers from Keidanren, Toyota, and Hitachi in Tokyo, 2011.

[13] UN Climate Change, Kyoto Protocol, status of ratification, https://unfccc.int/process/the-kyoto-protocol/status-of-ratification.

The high-profile conference also offered a rare opportunity for Japan to gain global prominence for international leadership on the issue.

From that point forward, environmental NGO professionals were more actively included in Japan's policymaking related to domestic as well as foreign policy. For example, the Institute for Global Environmental Strategies (IGES) was established in 1998 to promote research cooperation on climate-related issues, with a particular focus on the Asian region. It regularly hosts both high-level and working-level workshops on specific environmental issues prioritized by the government (everything from waste management to climate change mitigation) while also providing technical and policy advice to officials at all levels of government (Haddad 2021).

Japan's foreign policy began to prioritize sustainable development in the projects it supported abroad and advocated for increased attentiveness to environmental pollution and climate change in international venues (Okano-Heijmans 2012; Schreurs 2005). Additionally, its environmental lawyers became more active internationally, promoting the idea that environmental rights are human rights (Avenell 2017).

Second, 1997 was the year that Toyota launched its first mass-market hybrid vehicle, the Prius. Japanese automakers had been working on high-efficiency, electric, and hybrid vehicles for many years, but the success of the Prius was different. As a Toyota manager explained to me during a 2011 interview in Tokyo:

> The first-generation Prius didn't have much commercial success, but Toyota as an environmental brand image jumped a lot, way ahead of Honda or Nissan. That was a real turning point. Pollution problems may have started it, but we saw that the environment can improve the competitiveness of the company. It can enhance the brand and the company. That aspect had not been commonly understood – the environment can be commercially useful.

Over the next twenty years, Toyota's pro-environment policies paid off: in 2022 (most recent data available), Toyota had the largest share (11.5 percent) of the global auto market, more than four times the largest US manufacturer (Ford and Chevrolet both had 2.8 percent).[14]

Finally, three disasters (the global financial crisis in 2008–9; the earthquake–tsunami–nuclear disaster in 2011; and the COVID pandemic in 2020–22) caused massive disruptions to the Japanese economy, and the government's decision to prioritize "green growth" in all three recovery strategies gave large boosts to its eco-developmental state. The Lehman Shock, as the 2008–9 global

---

[14] "Global Automotive Market Share in 2022, by Brand," Statista, www.statista.com/statistics/316786/global-market-share-of-the-leading-automakers/.

financial crisis is commonly called in Japan, impacted Japan particularly hard, resulting in the collapse of its stock market, exports, and manufacturing (Kawai and Takagi 2011; Sommer 2009). Like other countries, Japan saw renewable energy and other "green growth" technologies as key to national and global recovery (JETRO 2009; OECD 2009).

Before it had fully recovered from the global financial crisis, Japan was hit with a crisis of a different kind. On March 11, 2011, the most powerful earthquake ever recorded in Japan triggered a devastating tsunami (up to 40 meters high in some places), which in turn caused a meltdown at the Fukushima nuclear facility. Nearly 20,000 people died, hundreds of thousands were displaced, and the World Bank estimated that the cost of the disaster exceeded $200 billion USD.[15] Global supply chains were highly disrupted (Fisher 2011), and business, citizens, and governments abruptly recognized the need for a dramatic reduction in energy consumption and a dramatic increase in renewable energy. Thus, the earthquake–tsunami–nuclear devastation, which reduced coastal towns to rubble and wiped out acres of farmland, created both the opportunity and the imperative for the rapid expansion of renewable energy and green development across the islands (Fraser 2020). In the decade since the Fukushima disaster, Japan has enhanced its commitments to the UN Sustainable Development Goals (SDGs) and expanded its development of green finance mechanisms (Elder and King 2018).

Not quite a decade later, the COVID-19 global pandemic hit and was more catastrophic to the Japanese economy than the previous two crises combined. Although Japan fared comparatively well during the initial crisis in 2020, quickly locking down its elderly and masking up (Tiberghien 2021), it had more difficulty coping with the highly transmissible newer variants and struggled with vaccination (Roberts and Kelman 2022). By this time, however, Japan's eco-environmental state systems were well in place, so its gigantic recovery package included large and highly ambitious environmental policy goals and significant government funding. Following the lead of China's President Xi, Japan's Prime Minister Yoshihide Suga announced Japan's commitment to reaching Net Zero emissions by 2050 in October 2020. The following summer, the 2020 Olympics (held in the summer of 2021) showcased many of Japan's green technologies, including powering the games using hydrogen, reusing or recycling 99 percent of the nonconsumable items procured for the games, making all of the medals from materials extracted from discarded electronics, and promoting accessibility and inclusion through its #WeThe15

---

[15] Wikipedia has an excellent overview of the disaster: "2011 Tōhoku Earthquake and Tsunami," Wikipedia, https://en.wikipedia.org/wiki/2011_T%C5%8Dhoku_earthquake_and_tsunami.

campaign to end discrimination against people with disabilities.[16] That fall, the Ministry of Economy, Trade and Industry (METI) hosted Tokyo "Beyond Zero" Week 2021, which comprised six global conferences related to the environment, including the first ministerial meeting of the Asia Green Growth Partnership as well as conferences focused on carbon recycling, hydrogen, ammonia, and other innovative energy technologies.[17]

Today, Japan's eco-developmental state is fully consolidated and is making significant headway in promoting environmental policy and practices across its archipelago, in the broader Asian region, and throughout the world. It is investing heavily in a number of critical technologies – in April 2022 announcing a 2 trillion yen ($15 billion USD) Green Innovation Fund that will direct resources to accelerate development and promote national competitive advantage in emerging technologies such as hydrogen power, green buildings, sustainable lifestyle technologies, and more.[18] Policymakers are working closely with business, NPOs, and other stakeholders in designing and implementing policies that address climate change and improve environmental outcomes across a wide range of policy fields, from technology, to education, to labor relations. Finally, the Japanese public continues to support its government; the LDP has retained a commanding majority of the national legislature (56 percent of seats in the most recent general election). Environmental activists and volunteers continue to make positive changes in their own communities, support pro-environmental policy change at the national level, and collaborate with activists around the world to spread positive action internationally.

3.2 South Korea

Environmental movements in South Korea, like in Japan, began as local, grassroots protests against industrial pollution that was threatening the lives and livelihoods of people living near large industrial complexes. As in Japan, initial protests were met with repression as political and corporate actors sought to quash resistance to industrial expansion and eliminate any questioning of the country's growth-first policies. The movements began in the 1970s and expanded in the 1980s while the country was governed by nationalist/military parties. Unlike the early Japanese protesters, however, South Koreans had the benefit of being able to learn from environmental and civil rights movements in

---

[16] #WeThe15 is a movement to "transform the lives of the world's 1.2 billion persons with disabilities who represent 15% of the global population," www.wethe15.org/.

[17] METI, Tokyo "Beyond Zero" Week, www.meti.go.jp/english/policy/energy_environment/global_warming/roadmap/.

[18] New Energy and Industrial Technology Development Organization (NEDO), "Overview of the Green Innovation Fund Projects," https://green-innovation.nedo.go.jp/en/about/.

other countries. They leveraged the organizational and political experience of movements in other countries to gain political traction, linking their environmental causes to pro-democratic agendas (Haddad 2015; Ku 2011).

In South Korea, national environmental movements were sparked by high-profile industrial pollution cases such as the one in Ulsan that galvanized locals. Their plight was taken up by opposition parties who used their cause to engage the rest of the country (Ku 2002). Support for the anti-pollution movement spread, ultimately merging with the pro-democracy movements as people recognized that the roots of the environmental problems were political: leaders were not prioritizing the needs of the people over the interests of corporations.

Social movements expanded across the peninsula as part of a broader effort to extend participatory democracy, which ultimately resulted in the (relatively) peaceful democratization of the country. The Korea Federation for Environmental Movements (KFEM) grew to become the region's largest environmental organization, and it has continued to support liberal parties and political leaders in national and local elections (Ku 2011; Lee 2000).

The frequency and intensity of climate-related disasters increased simultaneously with the growing understanding among the public and policymakers about the links between natural disasters and human-induced climate change. As a result, environmental advocacy shifted from reactive, anti-pollution policies to proactive environmental policies. Climate-related natural disasters helped galvanize the public on the importance of environmental and climate issues. In particular, in 2002, just ten years after the first Earth Summit, South Korea was hit by a devastating typhoon (Rusa), and a year later it was hit with another one (Maemi). The economic impact of Maemi was particularly large (more than $500 million USD) because it hit the industrial port city of Busan.[19]

The global economic crisis of 2008–9, and especially the subsequent giant fiscal stimulus packages focused on developing green industries, served as an inflection point that decisively shifted East Asian developmental states to become eco-developmental states. The South Korean case demonstrates that these processes are not inevitable. The personal commitment, vision, and leadership of a single individual at a critical moment catalyzed national and regional transformation.

In the early 2000s, just before the global financial crisis, Lee Myung-bak was the mayor of Seoul. His signature project was the restoration of the Cheonggyecheon river, which ran through the center of the city and was covered by an ugly highway when he took office. Through a massive public–private

---

[19] "Managing Typhoon Risk in South Korea," AIR Worldwide, December 14, 2010, www.air-worldwide.com/publications/air-currents/2010/Managing-Typhoon-Risk-in-South-Korea/.

collaboration, the highway was dismantled, the river landscaped, and a vibrant public park was created in the center of the city. The project was widely seen as a success both by the residents of Seoul and by urban planners around the world (Cho 2010; Lee and Anderson 2013), winning the Urban Transport Award in 2006, the Veronica Rudge Green Prize in Urban Design in 2010, and the INDEX Design to Improve Life Award in 2011.

The project's success helped Lee Myung-bak win the presidency in South Korea's 2007 election; nearly half of South Korea's population lives in Seoul and would have had access to the park and been able to witness its transformation. As a businessman (he had been Hyundai's youngest CEO), his approach to public policy in general and environmental issues in particular was decidedly pro-business. In 2008, during the sixtieth anniversary of the founding of the Republic of Korea, President Lee announced his "low carbon, green growth" strategy, which aimed to pivot Korea's economy toward the development of low-carbon industries and technologies as a way of enhancing the country's international competitive advantage (Republic of Korea 2011).

When the global financial crisis hit a few months later, his green growth vision expanded dramatically, eventually turning into a National Strategy for Green Growth that invested more than 100 trillion won ($87 billion USD) into a wide range of environmental projects from adaptation to R & D for renewable energy development (World Bank 2021). South Korea has continued to champion green growth on the global stage, and Seoul won the global competition to host the secretariat of the Global Green Growth Institute, which was established at the Rio+20 conference in 2012 and charged with administering the Green Growth Fund, a global mechanism to finance sustainable development projects around the world.

Commitment to green growth as a mechanism for both economic development and social transformation has only intensified in the wake of the COVID-19 pandemic. President Moon Jae-in made South Korea's Green New Deal a cornerstone of the Democratic Party's 2020 parliamentary elections as a national strategy to revitalize the economy after the pandemic (Lee and Woo 2020). Korea's commitment to green growth as a means of bolstering the country's competitive advantage has continued even when the conservatives regained control of the government in 2022. At a 2023 speech to the World Economic Forum, President Yoon touted Korea's hydrogen and battery technology as well as its commitment to green official development assistance (ODA) to support sustainable prosperity around the world.

Korea's use of green growth as a focus of its national economic and foreign policies by administrations from opposing political parties is an indication of the maturation of its eco-developmental state. Developing green industries and services has become a cornerstone of its industrial policy. Business actively collaborates with the government in making policy and includes a diversity of voices in policymaking. For example, President Yoon's cabinet includes a start-up entrepreneur (Lee Young), an NGO professional (Ha Wha-jin), and a journalist (Park Bo-gyoon). Finally, the citizens of South Korea continue to support their government: the most recent World Values Survey found that 51 percent of South Koreans have "a great deal" or "quite a lot" of confidence in their government (in contrast to just 33 percent of people in the United States who voice the same level of confidence in their own government).[20]

3.3 Taiwan

As was the case in Japan and South Korea, Taiwan's environmental politics are rooted in public opposition to industrial pollution. In the mid-1980s, Taiwanese started to push back against the growth-at-any-cost policies of their government, filing lawsuits and staging public protests. The case that garnered the most notoriety came to be known as the Lukang Rebellion. Villagers in Lukang mounted an increasingly sophisticated resistance effort to a proposed DuPont facility, including local listening sessions, lawsuits, public protests, and a public relations campaign. In the end, the locals won and DuPont pulled out (Reardon-Anderson 1997). Buoyed by their success, environmental activists across Taiwan teamed up with pro-democracy activists to repeal martial law and democratize (Hsiao 1999; Tang and Tang 1997).

After democratization, Taiwanese environmental movements aligned themselves with liberal parties and commonly engaged in protest politics to raise the profile of their issues, which were often catalyzed by NIMBY protests against new or expanding industrial parks (Ho 2014). Unlike other places in Asia, Taiwan's Green Party took root, moving some of the environmental political battles into local legislatures. Politicians from the Green Party were elected to local assemblies and gained prominent positions in municipal governments, including in Taipei (Fell 2021; Grano 2015).

Similar to South Korea, Taiwan saw its environmental politics shift from reactive to proactive in the early 2000s. Taiwan's environmental campaigns have evolved to become highly sophisticated national efforts that mobilize large cross-sections of society and use a variety of tactics, including public protests, academic study groups, and celebrity social media campaigns (Fell 2017; Ho 2014). Additionally, Tu Wen-Ling has documented that citizens are using new technology to generate citizen-science that contests official information about

---

[20] World Values Survey Online data, www.worldvaluessurvey.org/WVSOnline.jsp.

pollution and pressures policymakers to improve their air quality management measures (Tu 2019).

Rising climate threats have also contributed to the public's increased awareness of the threats posed by climate change. In 2000, the "super-typhoon" Bilis devastated the main island, forcing thousands to evacuate and causing widespread mudslides.[21] In 2009, Typhoon Morakot killed more than 500 people,[22] and Taiwan experienced its worst drought in more than 50 years in 2021, drying up the iconic Sun Moon Lake, causing power shortages, and slowing production in its semiconductor chip industry.[23]

Like South Korea's Lee Myung-bak, Chen Shui-bian was able to leverage his success as mayor of the capital city to generate the support necessary to become president of Taiwan in 2000. While he was mayor of Taipei, Chen Shui-bian instituted the city's famous "keep trash off the ground" campaign.[24] His successor, the conservative Ma Ying-jeou, who followed Chen first as mayor of Taipei and then as president (2008–16), added a "pay as you throw" component to the municipal waste in Taipei, requiring residents to use special bags for disposing of garbage and ultimately ending the use of landfills for city waste (Chiu 2002).

Ultimately, Ma became a president who championed Taiwan's environmental industries as a path of economic recovery and building competitive advantage. As a conservative leader who was president at the time of the global financial crisis, Ma Ying-jeou followed South Korea's lead and focused on "green growth" as a key strategy for economic recovery, arguing that "a country's green competitive edge will decide its position on the world stage 10 or 15 years later."[25] In 2011, he launched the Green Trade Promotion Program as a cornerstone of the effort to make Taiwan a hub of development of green industries, including green energy, smart cities, transportation, and circular economy.[26] Ma's successor, President Tsai Ing-wen, has only increased the government's commitment to environmental policy goals, including funding

---

[21] NASA/Jet Propulsion Laboratory, "Super Typhoon Bilis Hurls Rain and Wind," *ScienceDaily*, August 24, 2000, www.sciencedaily.com/releases/2000/08/000824082159.htm.

[22] "TIMELINE: Major Typhoons to Hit Taiwan," *Reuters*, August 14, 2009, www.reuters.com/article/idUSB479427.

[23] "Taiwan Prays for Rain and Scrambles to Save Water," *New York Times*, May 28, 2021, www.nytimes.com/2021/05/28/world/asia/taiwan-drought.html.

[24] This *New York Times* article about the trash collection system in Taipei includes a video of how it works: "Taiwan Dispatch: When You Hear Beethoven, It's Time to Take Out the Trash (and Mingle)," *New York Times*, February 8, www.nytimes.com/2022/02/08/world/asia/taiwan-waste-management-beethoven.html.

[25] "Ma Encourages Firms to Go Green," *Taipei Times*, May 23, 2010, www.taipeitimes.com/News/taiwan/archives/2010/05/23/2003473669.

[26] Green Trade Promotion Office, www.greentrade.org.tw/en/aboutus/about-us.

for renewable energy and other green technologies in the wake of the COVID-19 crisis.[27]

A critical component of President Tsai's environmental policy, which remains quite distinct from those of previous Taiwanese leaders as well as other leaders in the region, is the extent to which she has raised up and supported Indigenous peoples as important collaborators in the collective effort to combat climate change. Indigenous groups have been playing an increasingly large role in Taiwan's environmental movement (Chang 2020; Fan 2021; Nedopil 2021), and President Tsai actively partners with Indigenous groups to combat climate change and promote green businesses (Nedopil 2021).

In sum, as with South Korea, across multiple administrations Taiwan's leaders have continued to emphasize the importance of green technology as key to developing global competitive advantage, indicating a maturation of its eco-developmental state. Business continues to have a strong role in policy-making, while diverse voices from the nonprofit and academic communities are also included (Grano 2020). Finally, although Taiwan's population struggles with numerous political challenges, not least of which are military threats from the mainland, it generally supports and trusts its government even if it may punish political parties at the ballot box. Like South Korea, Taiwanese have a high level of confidence in their government (52 percent had "a great deal" or "quite a lot" of confidence in their government according to the most recent World Values Survey).[28]

3.4 China

Although the form that environmental politics has taken in the People's Republic of China has been quite different than that experienced in democratic Japan, South Korea, and Taiwan, the general trajectory of public pressure followed by corporate and governmental policy change has been remarkably similar. Whereas Japan's period of rapid industrialization happened in the 1950s and 1960s, and South Korea's and Taiwan's happened in the 1970s and 1980s, China's took place during the 1990s and 2000s. China's accession to the WTO in 2001 was a turning point, opening the country up to international investment and ultra-rapid industrialization. What had been piecemeal open trade policies through free trade zones gradually opened the entire country up to foreign investment and rapid industrialization. Foreign investment was (and is) always done through cooperation with local actors and in collaboration with

---

[27] Office of the President, Earth Day 2022, https://english.president.gov.tw/NEWS/6263.

[28] World Values Survey (2017–22), Confidence in Government (Taiwan ROC): www.worldvaluessurvey.org/WVSOnline.jsp.

government, allowing the government and the Chinese Communist Party (CCP) to direct resources and control the flow (Gallagher 2002; Rawski 1999).

In China, three concurrent trends in the late 1980s and early 1990s together helped create both the political imperative and the economic opportunity to integrate anti-pollution (although not yet pro-climate) measures into policy-making. First, as in the other East Asian countries, rapid industrialization led to intense industrial pollution and local protests erupted, focusing on the devastating health and livelihood consequences of pollution and their links to political corruption. China's protests were bigger, more frequent, and more widespread than anywhere else in the region – official statistics reported tens of thousands of environment-related protests every year during that period (Zissis and Bajoria 2008). China's public protests came to a head in 1989, when pro-democracy protests in Tiananmen Square created a crisis within the CCP. Fed up with corruption and rising inequality, students and others in the square were inspired by the successful pro-democracy movements in neighboring South Korea, Taiwan, and the Philippines and by the Eastern European activism that would lead to the Velvet Revolution and other transitions.

The CCP wanted to avoid following the path taken by ruling parties in South Korea and Taiwan, which ultimately led to the democratization of those countries and their loss of political power. Instead, it sought to follow the route taken by the LDP in Japan, which was able to stay in power by rapidly accommodating citizens' demands for greater environmental protections. Six months after protesters were cleared from Tiananmen Square, China enacted a significantly enhanced Environmental Protection Law and quickly followed up with numerous additional laws aimed at curbing air, water, solid waste, and noise pollution (Xie 2020).

Concurrent with domestic pressure to address pollution, the global environmental movement was gaining momentum, and these organizations realized that China would play a critical role in the success (or failure) of global efforts to address climate change. Global environmental organizations began establishing offices in China (e.g., the World Wildlife Fund [WWF] entered China in 1980,[29] Friends of the Earth [FoE] opened an office in Hong Kong in 1983 and started working on the mainland in 1992,[30] and the Natural Resources Defense Council began work in China in the mid-1990s,[31] to name a few), working with Chinese government officials and businesses to develop pro-environmental policies for

---

[29] WWF China History, https://en.wwfchina.org/en/who_we_are/our_history/.

[30] China Development Brief, Friends of Nature, https://chinadevelopmentbrief.org/ngos/friends-of-nature-2/.

[31] Natural Resources Defense Council, China, www.nrdc.cn/aboutus?cid=11&cook=1#xm_fruit&cook=1.

the Chinese context and spreading environmental awareness among the Chinese public. Their expertise and funding helped the Chinese government leverage international resources in their transition to a less-polluting economic growth model (Economy 2004).

Finally, a third important trend in China in the 1990s was the development and spread of internationally recognized environmental management standards. The most important of these emerged in 1996 when the ISO introduced its ISO 14000 series of standards,[32] which gave companies clear guidance on creating comprehensive environmental management within their organizations and a method to certify that this was being followed. Many multinational corporations began requiring ISO certification for their suppliers, many suppliers used ISO certification to boost their attractiveness to global customers, and local as well as national government officials actively encouraged Chinese firms to obtain certification as a way to boost commercial competitiveness (Li 2019). The role of ISO certification in promoting pro-environmental business practices is discussed in greater detail in Section 4.1.

Thus, at the start of the twenty-first century, China had made its first policy shift away from growth-at-any-cost industrial policy toward policies aimed at curtailing the negative health impacts of industrial pollution. However, its policies did not yet systematically address the broader issue of climate change. The shift to pro-climate policy began during the late 2000s, boosted by the 2008 Beijing Olympics, which served as an important focal point for pro-environment and pro-climate activists in civil society, business, and the government. Following Sydney's example, Beijing made environmental sustainability a key component of its bid for the 2008 games, which it would quickly brand as the Green Olympics. From 2001, when it won the bid, until the games themselves in 2008, the city quickly ramped up its efforts to clean its air and water, planting trees, reducing car and power plant emissions, and expanding public transportation and green building construction (Turner and Ellis 2007).

Immediately after the Olympics, the global financial crisis of 2008–9 created a system-wide shock to China's model of economic growth, which had relied heavily on export-oriented industries and migrant labor (Lardy and Subramanian 2011). Like other countries in the region, China responded with a very large stimulus package that prioritized green investment – especially with respect to power generation, transportation, and other green technologies (Cai et al. 2011), as well as rural development (Naughton 2009).

---

[32] You can find out more about the ISO 14000 series of standards from the ISO website, www .iso.org/iso-14001-environmental-management.html.

The new investment and governmental focus on better environmental performance (environmental performance was added to the cadre evaluation system in 2006) (Wang 2013) combined with the spread of social media to pressure and reward public and private actors for behaving in pro-environmental ways. In 2015, China's President Xi Jinping seized the opportunity to take a global leadership role, joining with US President Barack Obama and other leaders to sign the Paris Agreement.

At that time, in 2015, China had all the hallmarks of a fully consolidated eco-developmental state. It was rapidly expanding national investment in renewable energy, electric vehicles, organic agriculture, and other pro-environmental products and services that would gain it a competitive advantage in world markets. Policymakers were actively seeking input from not only large state-owned and private companies but also global NGO professionals, academics, and local activists who could help inform China's future policy directions. Additionally, although some citizens were unhappy about restrictions on free speech and incidents of political repression, the government appeared to have the widespread support of its people.

Initially, the positive trends in terms of environmental policy appeared to continue. In 2017, after US President Trump withdrew from the Paris Agreement, President Xi did not back away from his environmental commitments but rather stepped forward, claiming the role of global leader on climate.[33] Then, at the celebratory seventy-fifth meeting of the UN General Assembly in September 2020, President Xi enhanced that leadership position by committing his country to Net Zero by 2060, a month before Japan's Prime Minister Suga and South Korea's President Moon made their own commitments to become Net Zero by 2050 (the United States followed six months later in April 2021).

However, in 2016, the year after the Paris Agreement was signed and before President Trump withdrew the United States, China passed a new Charity Law that clarified the legal status of social and NPOs in China, enabling some additional legal avenues for incorporation while restricting other activities (Lin and Zhou 2022). It was followed the next year with the Overseas NGO Law that significantly curtailed the ability of international NGOs to operate in China (Shieh 2018). The two laws together have had a chilling effect on China's civil society, making it much more difficult for the NGO sector to nurture the kinds of professionals able to offer policymakers the technical and political advice necessary to make eco-developmental states function (Holbig and Lang 2022; Sidel 2019; Spires 2020).

---

[33] "Is China Really Stepping Up As the World's New Climate Leader?," *The World*, November 8, 2017, www.pri.org/stories/2017-11-08/china-really-stepping-world-s-new-climate-leader.

China's classification as a developmental state has always been somewhat controversial (Pearson, Rithmire, and Tsai 2021, 2023; Zhang 2018; Zheng 2022), and it is the only government among the four that is not a democracy. Since 2018, when China's National People's Congress removed the two-term limit for presidents, political tightening has intensified. In the last few years, the space for civic engagement, including environmental activism, has become progressively more restricted, while political repression has greatly increased (Göbel 2021; Liu 2020).

Furthermore, the third pillar of developmental states – widespread support from the public – has been significantly challenged by the COVID-19 pandemic and the subsequent government response. When the health crisis first began in early 2020, the Chinese public strongly supported their government's "zero-COVID" policies, which were discussed by Chinese (and others) as responsible, science-driven, and publicly minded. China's response was viewed in stark contrast to the erratic, hostile, and suspicious view that people in the United States had of their own government's policies (Luo et al. 2021; Xing, Li, and Wang 2021). However, by late 2022, the Chinese public's view shifted as the pandemic wore on, lockdowns lasted longer, and punitive responses to violations became harsher. Furthermore, the "blank sheet" protests, as the widespread protests in December 2022 across China came to be called, were about more than just zero-COVID policies, as citizens took the opportunity to indicate broader dissatisfaction with censorship, political repression, and other, unspecified, policies (Murphy 2022). A few weeks after the "blank sheet" protests began, the government abruptly ended its "zero-COVID" policy, indicating that it was attentive to the public's concerns. Remaining protesters were jailed, and the protests dissipated.

As of this writing, it is too early to determine if these departures from China's eco-developmental state policymaking will persist. China remains committed to supporting the development and expansion of green technology and finance (discussed at greater length in Section 4). What is not clear is whether it will continue to include and expand nongovernmental voices in policymaking and whether it will retain the broad support of its citizens. If the Chinese government reprioritizes its environmental goals and includes more perspectives in its decision-making, it may regain the support of its people. If that happens, then the events of late 2022 will likely be viewed as a short setback for a broader pattern in which China's eco-developmental state evolves and grows stronger. However, if the country's pursuit of economic competitive advantage supplants environmental concerns, if it further restricts the diversity of voices heard in policymaking, and if public support continues to erode, then the brief period after the global financial crisis and before the COVID pandemic will be viewed as one where China experimented with an eco-developmental state model but ultimately rejected it. At the time of this writing, it is too soon to determine which path China will take.

3.5 Conclusion

East Asian countries have followed a similar environmental policymaking trajectory. They all began with industrial policies focused on rapid industrialization and economic growth with little concern for the environmental consequences. First in Japan (in the 1890s and then again in the 1960s), then in South Korea and Taiwan (1980s), and finally in China (2000–2010s), governments reacted to strong pressure from their public to enact stronger environmental regulations that would mitigate the negative effects of pollution on the lives and livelihoods of their citizens. In the early years of the twenty-first century, all four places dramatically expanded their pro-environmental policymaking as both the risks and the consequences of climate change grew and businesses found more ways to profit from greener products and processes.

At the time of this writing, Japan, South Korea, and Taiwan have fully consolidated and are further reinforcing all three pillars of their eco-developmental states: (1) they have government policies focused on promoting environmental products and services as a strategy for boosting national competitive advantage; (2) they have developed and expanded policymaking processes that include diverse perspectives; and (3) although the parties in power may change, governments in all three countries enjoy broad public support.

The situation is less clear in China. While it appears that the government is still committed to investing in pro-environmental industries as a way of enhancing its international competitive advantage, there has been a significant reduction in the diversity of voices in policymaking and the public's support of the government has eroded. It may be that the restriction of voices in policymaking and the loss of public support prove to be temporary, and China may soon return to its previously successful eco-developmental state model of governance. Alternatively, China's turn toward authoritarianism could intensify, further distancing itself from the economic, political, and environmental successes of its neighbors.

4 Green Business: Technological Innovation and Green Finance Creating Win-Win-Win-Win Solutions

Throughout this Element, we have seen East Asian governments frequently seeking pro-business solutions to environmental challenges. This section takes a deeper dive into that area of environmental politics, looking more closely at the role of green technology and green finance in promoting pro-environmental policies that create benefits for business, government, society, and the planet.

This section will begin with a discussion of why the eco-developmental states of East Asia are so good at pro-business environmental policymaking, and why it can be so effective. It will then use two specific types of pro-business environmental policymaking – green technology and green finance – to illustrate how pro-business eco-developmental policies can address many policy goals at once, promoting not only economic growth for companies but more and better jobs for workers, rural revitalization, cleaner air, and reduced carbon emissions.

4.1 Business Takes the Lead

Section 3 described the historical origins of East Asia's environmental politics, showing how rapid industrialization led to high pollution, which in turn led to strong citizen protests against pollution. The initial political and corporate responses were reactive – companies and governments crafted policies intended to placate angry residents. What is interesting is that their efforts did not stop with mild stopgap measures designed to offer minimum concessions to residents while still supporting polluting industries. Instead, East Asia's governments modified their pro-growth developmental state policymaking to include pro-environmental priorities and voices. For business, solving environmental problems became a market opportunity.

In the United States, the environmental movement was led by grassroots activists and well-funded professional NGOs who sued polluting companies and negligent governments, bought land for permanent conservation, organized rallies for environmental causes, and funded pro-environment politicians (Bosso 2005; Eisner 2006). In Europe, it was the government sector that pushed the agenda. Green parties joined national and local governments, inserting environmental concerns into legislation. Bureaucrats at local and national levels developed policies and disseminated them across the continent (Dalton 1994; Kijek 2015; O'Neill 1997).

In East Asia, it was often the business sector that pushed for stronger environmental policies. There are a few reasons why East Asia's business sector has been particularly willing to engage in pro-environmental innovations and press their governments for better environmental policies. First, the business context in East Asia allows for longer time horizons for investment payoffs, so costly pro-environmental investments that take longer to generate a positive return become possible. Second, East Asia's export-oriented economies and shifting consumer markets increasingly rewarded companies that could document and market green products and processes. Finally, by bringing connected stakeholders together and prioritizing collaborative decision-making, the eco-developmental state policymaking structure frequently generated solutions that offered benefits to many stakeholders, further supporting a shift toward innovative and proactive environmental policymaking.

Several features of the business context in East Asia promote the kind of longer-term capital investments that are needed for pro-environmental improvements. First, East Asian firms tend to be less reliant on stock market financing, which rewards quick value to shareholders, whereas bank or other forms of institutional financing generally allows for longer time horizons to demonstrate a return on investments (Hoshi, Kashyap, and Scharfstein 1991). Japanese, South Korean, Taiwanese, and Chinese firms all receive most of their funding from banks, institutional cross-shareholding, and government funding, so they have a lower reliance on capital markets for their financing compared to US or European firms (Haggard 1996; Noble and Ravenhill 2000).

Pro-environmental investments, such as installing solar panels on a warehouse rooftop, purchasing a fleet of electric vehicles, or building new office complexes heated with geothermal systems, require high initial capital expenditure that then pays off after several years. Over the lifetime of the building or vehicles, companies make back their initial investment several times over. For patient investors like banks and families, waiting many years for high positive returns makes financial sense. As a result, East Asian firms, which are less reliant on stock market financing, are able to wait longer for their investments to return a profit compared to counterparts elsewhere, making it easier to make environment-related investments (Lin, Yang, and Liou 2009; Nakamura 2011).

Another feature of the business context in East Asia that favors long-term investments is the region's tendency to have long-term relationships with their employees, suppliers, and customers. Although the model has faded somewhat in recent years, Japan's postwar political economy was based on a "lifetime employment system" where employees were hired immediately after they graduated from school and worked their entire lives for the same firm (Dore 1973; Estevez-Abe 2008). While South Korea does not have the same lifetime employment system, the dominant position of its chaebol has exerted a similar effect on its employment and social welfare systems (Bamber and Leggett 2001; Goodman and Peng 1996). Historically, China's corporate commitment to its employees, customers, and suppliers was even more stable than in Japan or South Korea, because state-owned enterprises, which guaranteed employment as well as housing and social services, dominated the commercial sector for decades (Aivazian, Ge, and Qiu 2005; Hassard et al. 2010). These long-term relationships provided stability in labor as well as supplier and distributor relationships, so firms could negotiate the complex arrangements needed to green their products and supply chains.

Finally, a high proportion of East Asian's biggest firms are family-owned. To give just two well-known examples, Akio Toyoda, the current president of Toyota Motor Corporation, is the great-grandson of Sakichi Toyoda, who founded the Toyoda family companies. The current chairman and CEO of the South Korean technology giant LG is Koo Kwang-mo, the grandson of LG's founder, Koo In-hwoi.

Like the finance and employment structures discussed in this section, family ownership frequently extends the time horizon for investments to pay off. The CEOs of family-owned corporations frequently see their role as one of steward-ship, maintaining or increasing the value of the company before handing it down to the next generation. When a business is family-owned, the reputation of the firm is tied up with that of the family. Negative press about environmental disasters taints the family's image as well as that of the company, just as positive news can enhance a family's reputation as well as their company's. Finally, family-owned firms often have more personal relationships with their employ-ees, suppliers, and customers, so they can be more willing to invest in preserv-ing, enhancing, and expanding those relationships than other types of businesses (The Economist 2014; Yeh, Lee, and Woidtke 2001).

In addition to corporate finance and ownership structures that support the long-term investments necessary for pro-environmental policies, East Asia's export-oriented economies have also facilitated prioritizing environmental investments. As discussed in Section 1, East Asia's developmental states generated their rapid economic growth by manufacturing goods for North American and European markets (Berger and Dore 1996; Haggard 1990). By the 1970s, both of those markets began to shift as governments set higher environmental standards for products in their markets. East Asian firms quickly adjusted, developing products that met or exceeded the environmental require-ments of their target markets. As the firms grew in size and global reach, it made economic sense for them to produce all products using the highest standard, since complex supply chains made it costly to produce similar products using different product specifications (Busch, Jörgens, and Tews 2005; Epstein and Roy 1998).

With the improvement in communication technologies, journalists and NGOs increasingly targeted global brands for investigation. When investigators dis-covered that multinationals, their subcontractors, or their subsidiaries were abusing workers, selling harmful products, or polluting the environment, activ-ists would hammer the companies across numerous global media platforms, costing the brands millions of dollars in lost sales and requiring millions more to restore consumer trust (Haddad 2015; IPE and NRDC 2014; Kraft, Stephan, and Abel 2011; Murdie and Urpelainen 2015).

Responding to consumer and regulator demands for proof of compliance, companies began to work on harmonizing environmental standards. The ISO is an international organization that sets international standards for a wide range of technical and manufacturing processes and establishes a method for firms to certify that they have met these standards. In 1996, it established the ISO 14000 series of standards focused on different aspects of environmental management (e.g., 14001 is an environmental management standard, 14040 is a standard for life-cycle assessment, etc.). These internationally recognized standards turned out to be a game-changer for green businesses because they allowed firms to demonstrate to investors that they were meeting internationally recognized standards.

East Asian firms have been among the largest participants in the ISO system. In 2021, the ISO issued ISO 14001 certificates to more than half a million different sites in 194 countries. China had by far the most certificates of any country, with 217,592, and Japan had the second largest with 21,976. South Korea and Taiwan had 6,886 and 2,614 environment-related certifications issued in the two countries respectively. The United States had comparatively few, with only 4,171.[34]

As the region prospered, East Asia became not just a base of production but also a valuable market. Firms that had initially targeted foreign markets turned toward the domestic Chinese market and discovered themselves at a commercial disadvantage compared to domestic-only firms, since the latter did not need to meet European or North American production standards. As a result, large export-oriented corporations (both foreign and domestic) increasingly found themselves working with NGOs to pressure governments to raise local environmental standards. Local standards that were on par with those in Europe and North America benefited globally active firms since their production facilities and methods already met the higher standard (Tokunaga 1992; Vogel 1996). This market dynamic meant that, paradoxically, large multinational corporations frequently allied with global environmental NGOs in advocating for higher environmental standards in East Asia (Haddad 2021).

Finally, as discussed in Section 3, policymaking in East Asia's developmental states was made through networks of elite actors across business, government, and other sectors who coordinate policy to generate high economic growth for

---

[34] ISO 2021 Survey data, http://isotc.iso.org/livelink/livelink?func=ll&objId=18808772&objAction=browse&viewType=1.

the region's companies. So, when these commercial factors began to push businesses to make more environmental choices to improve their competitiveness and ensure their long-term profitability, they began to exercise their political influence to pressure policymakers to shift regulations in ways that supported pro-environmental business choices.

Furthermore, these pro-environmental business interests were emerging in conjunction with similar interests in civil society and among the foreign policy establishment. In South Korea, business and government leaders recognized that fossil fuel dependence represented an operational risk to corporations at precisely the same political moment that activists were demanding cleaner energy (Kim and Thurbon 2015). Dependence on foreign oil was a strong security concern for Japan and Taiwan as well, accelerating the shift toward renewable energy and promoting a regional ecosystem that supported green development (Grano 2015; Elder 2018). In all three places, voters and globally connected NGO sectors directed their representatives in elite policymaking circles, allies in local governments, and activists on the streets to prioritize climate and the environment (Avenell 2020; Grano 2020; Haddad 2021).

Of all four of the East Asian governments, China had done the most to develop its green business sector. In many ways, this is because its environmental crisis has been the most intense (Economy 2004; Zissis and Bajoria 2008), but it has also been the direct result of its consultative policymaking systems. Although it lacks the kind of electoral pressure that multiparty politics presents in democratic states, for many years its consultative system generated a process where multiple perspectives were included in policymaking in ways that facilitated positive-sum solutions (Gao and Teets 2021; Guttman et al. 2018) Relatedly, China's gigantic domestic market and connection to international markets meant that policies promoting green businesses at home also promoted green competitive advantage in global markets (Qiu et al. 2020; Zameer, Wang, and Yasmeen 2020).

More recently, however, the outlook for green business in China has become more uncertain. On the one hand, scholars have found that President Xi's anticorruption campaigns have improved green innovation, investment in renewable energy, energy efficiency, and environmental compliance (Chen et al. 2022; Nie, Zhou, and Feng 2022). On the other hand, the economic and political disruptions of the COVID-19 pandemic have slowed China's previously rapid renewable energy transition (Eroğlu 2021; Jia, Wen, and Lin 2021). In both cases, it is difficult to determine if these effects will be short-lived or enduring and whether the positive trends in innovation and manufacturing will outweigh the negative trends in the energy sector.

In sum, although it may first appear counterintuitive for the business sector in East Asia to have been a champion of pro-environmental policies, the funding and corporate structure of many businesses in the region gave them the longer investment time horizons needed to make environmental investments pay off. Additionally, as North American and Europe tightened their environmental regulations, East Asia's export-oriented firms were incentivized to raise their own standards to match or exceed those found in target markets. Finally, East Asia's eco-developmental state policymaking structures rely on coordinated and collaborative decision-making among government, business, academic, and other elites. Since policy innovations that were simultaneously pro-business and pro-environmental generated benefits for multiple stakeholders and contributed to the development of national competitive advantage, environmental considerations have been increasingly included in policymaking. The next two sections illustrate how pro-business environmental policy solutions can generate win-win-win-win outcomes in two specific areas of green business: green technology and green finance.

4.2 Green Technology

East Asian governments have directed significant government financial stimulus efforts toward the development and deployment of green technology, especially in clean energy and green transportation. First, in 2009–10, in the aftermath of the global financial crisis, and then again in 2020–21, in response to the COVID pandemic, governments funneled trillions of dollars into the green technology sector. Just as was the case with past developmental state industrial support, eco-developmental states frequently use "green growth" investments to accomplish more than one policy goal. In this case, they are generally seeking not only to improve their national competitive advantage and reduce emissions but also to facilitate rural revitalization, youth employment, public health improvements, and other public priorities.

East Asian countries have made enormous investments in clean energy technology, recognizing that the market for all aspects of renewable energy technology is global and rapidly expanding. In addition to a commitment to combat climate change, there is wide recognition that companies that develop renewable energy technology will gain access to enormous markets very quickly, and all countries in the region have prioritized these industries as a way of strengthening national competitive advantage.

Although the global energy transition is not as fast as many would like, it has been extremely rapid: in 2000, the installed capacity of solar power around the world was just 1 TWh. By 2020, it had jumped more than 800-fold to 856 TWh.

To put those numbers in perspective, New York City uses about 50 TWh of energy in a year.[35] So, in 2000 the world's total solar power was only one-fiftieth of New York City's needs, and by 2020 China could power New York City for more than four years on its solar power alone. Since China has dozens of megacities, it may seem like a drop in the ocean from a climate change perspective. That said, in 2020 China was producing 27 percent of its energy from renewable sources,[36] which was more than the total energy consumption of Japan, South Korea, and Taiwan combined.[37]

In addition to wind, solar, and geothermal power, there has been considerable regional investment in hydrogen. Japan in particular has been a global leader in this new field, declaring its plans to shift from a carbon-based society to a "Hydrogen Society" by 2050. Japan began investing in hydrogen-based fuel cells in the 1970s as part of its "sunshine" program that increased government investment in renewables and low- and zero-emission cars in response to the oil crisis. The current boom in hydrogen began after the country's nuclear power plants shut down after the 2011 disaster in Fukushima. Now public–private partnerships have created a number of "smart cities" where hydrogen heat and power units are being installed, and hydrogen refueling stations are proliferating to support the growing number of hydrogen vehicles on the road (Uriu 2021). South Korea announced its own Hydrogen Roadmap in 2018, China began promoting hydrogen vehicles in 2019 (Uriu 2021), and the Taiwanese government unveiled its "hydrogen roadmap" in 2022 (Biogradlija 2022).

Not surprisingly, the auto industry is a second huge area of investment, innovation, and growth for green technology in the region. China, Japan, and South Korea together make most of the cars in the world. In 2020, they produced almost 37 million vehicles (25 million in China alone),[38] and they are all moving very fast to develop and produce hybrid, electric, and hydrogen vehicles. Governments in the region have engaged in a wide array of policy measures and incentives designed to support the automakers in R & D as well as promote consumer demand by making electric vehicles cheaper and reduce concerns about insufficient charging stations or other factors related to convenience and customer satisfaction (Kim et al. 2019; Yarime and Karlsson 2018).

---

[35] New York Building Congress, *Electricity Outlook 2017: Powering New York City's Future*, www .buildingcongress.com/advocacy-and-reports/reports-and-analysis/Electricity-Outlook-2017-Powering-New-York-Citys-Future/The-Electricity-Outlook-to-2027.html#:~:text=Annual %20energy%20consumption%20in%20New,year%20over%20the%20forecast%20period.

[36] China Energy Portal, https://chinaenergyportal.org/2020-electricity-other-energy-statistics-preliminary/.

[37] Enerdata, "World Energy and Climate Statistics: Yearbook," https://yearbook.enerdata.net/total-energy/world-consumption-statistics.html.

[38] International Organization of Motor Vehicles Manufacturers (OICA), 2020 Production Statistics data, www.oica.net/category/production-statistics/2020-statistics/.

Because of their large carbon emissions and global markets, the energy and transportation sectors have been a particular focus for green technological development, but they are by no means the only sectors enjoying a green revolution. Addressing climate change requires a wholesale transformation of our carbon-intensive economy, and entrepreneurs across the region are finding creative ways to solve problems and meet consumer needs in greener ways in a wide array of different ways. The following four additional examples, one from each of the East Asian countries, provide a glimpse into the range of green tech innovation occurring in the region.

- Alipay Ant Forest is a mobile gaming app that encourages users to live a greener lifestyle by integrating eco-friendly activities, such as walking to work, into daily life. Users create "virtual green energy" through the Ant Forest platform, which eventually grows into an entire virtual tree. Once the virtual tree is fully grown, Alipay and its NGO partners plant a real tree in areas of northwest China at risk of desertification or commit to protecting conservation areas. From 2016, when it started, until 2019, the app had attracted more than 500 million users, planted 100 million trees, and generated 400,000 farm-related jobs in rural China.[39]
- Metro Farms is a ten-year collaboration between the Seoul Metropolitan Government, Seoul Metro, and Farm8, which broke ground in 2019. It is an AI-enhanced vertical smart farm established in previously unused parts of the Seoul Metro System, which creates jobs, grows local food, and purifies the air in the subway tunnels. Subway passengers can buy sandwiches and salads at the attached café or from the salad vending machines. The farm also offers educational tours for children and sells extra produce to local restaurants.[40]
- Kuradashi, launched in 2015 in Japan, is an e-commerce site aiming to reduce food waste by selling unsold food at a discount. Through a network of 800 companies selling more than 50,000 food products, the site reduces waste, prevents hunger, and benefits the planet. Membership more than doubled during the pandemic, rising from 80,000 to 180,000 between 2019 and 2021.[41]

---

[39] UN Climate Change, "Alipay Ant Forest: Using Digital Technologies to Scale Up Climate Action: China," https://unfccc.int/climate-action/momentum-for-change/planetary-health/alipay-ant-forest.

[40] Nikkei, "Underground Farms Sprout in Seoul's Subway Stations," January 14, 2020, https://asia.nikkei.com/Business/Startups/Underground-farms-sprout-in-Seoul-s-subway-stations.

[41] GreenBiz, "In Japan, Kuradashi Cuts Food Waste Creatively," July 28, 2016, www.greenbiz.com/article/japan-kuradashi-cuts-food-waste-creatively, www.reuters.com/article/us-japan-economy-foodwaste/japanese-companies-go-high-tech-in-the-battle-against-food-waste-idUSKCN2AS0RI.

- Umorfil bionic fibers, produced by Camangi Corporation in Taiwan, uses waste fish scales generated by Taiwan's fishing industry to create a new eco-friendly versions of petroleum-based textile materials such as nylon and polyester. The new fabrics are soft, naturally deodorizing, and 100 percent biodegradable.[42]

East Asia's efforts to promote green business may be led by the business sector, but it is also supported by a wide array of nonprofit and other actors who work with the government to promote pro-environment, pro-business policies across the region. One of the best examples is Ma Jun, who founded the Institute of Public and Environmental Affairs (IPE) in China in 2006. Originally an investigative journalist, Ma created a transparency-based platform that made it significantly easier for anyone to access public pollution data collected by the Chinese government. The organization then used the information to conduct public campaigns against global companies like Apple, who were sourcing from polluting subcontractors, encouraging them to clean up their supply chains. Ma told me in an interview that IPE's intent is not to shut companies down but to help them clean up. He and his organization worked with the Chinese government to launch the Blue Map App where users can report environmental violations directly to the government. The environmental records of corporations listed on the Hong Kong stock exchange are then listed, enabling banks and investors to reward companies with good environmental records (Haddad 2015, 2021).

East Asia's investment in green technology as a means of promoting national competitive advantage is paying off. This is most obvious in the case of China. In 2022, eight of the top ten solar panel manufacturers were Chinese,[43] it had 60 percent of the global market in electric vehicle sales (McKerracher 2022), and it has more high-speed rail than the rest of the world combined (Lawrence, Bullock, and Liu 2019). Japan, South Korea, and Taiwan are also benefiting from their countries' promotion of green tech: East Asia dominates the electric vehicle battery market – all ten of the top companies are Chinese, South Korean, or Japanese;[44] Japan and South Korea dominate the hydrogen passenger car market;[45] and all four

---

[42] Fashion United, "Six Sustainable Textile Innovations from Taiwan," November 13, 2020, https://fashionunited.uk/news/business/six-sustainable-textile-innovations-from-taiwan/2020111351938.

[43] Blackridge Research & Consulting, "Top 10 List of Solar PV Module Manufacturers in 2022," May 11, 2022 (blog), www.blackridgeresearch.com/blog/top-solar-pv-module-panel-manufacturers-companies-suppliers-producers.

[44] E-Vehicle Info, "Top 10 EV Battery Manufacturers in World by Market Share," https://e-vehicleinfo.com/global/ev-battery-manufacturers-in-world-by-market-share/.

[45] Interact Analysis, "Hydrogen Fuel Cell Vehicles in Japan and South Korea: Market Roll Out with Governmental Support (Part 1)," September 29, 2022, www.interactanalysis.com/hydrogen-fuel-cell-vehicles-in-japan-south-korea-market-roll-out-with-governmental-support/.

places are among the top ten photovoltaic exporting countries in the world (beating the United States, which has the number ten spot).[46]

In sum, green technology is a rapidly expanding part of East Asia's economy. Supported by a complex ecosystem of governments, nonprofits, and business entrepreneurs, new technologies in a wide range of fields are generating profits for businesses even as they reduce carbon emissions and enhance social and health benefits for people in the region. Because of its capacity to offer political benefits by pleasing both businesses and consumers while also addressing climate and pollution concerns of the NGO sector, it has become a focal point of government investment by East Asian governments.

4.3 Green Finance

"Green finance" is a general term used to describe financial tools (e.g., grants, loans, investments, insurance, etc.) that increase monetary flows to sustainable development priorities (Lindenberg 2014).[47] Nearly all forms of green finance have grown exponentially in the last few years. The 2020 UN *Financing for Sustainable Development Report* notes that investment strategies that consider the impact of environmental, social and governance (ESG) factors have "increased by 34 percent between 2016 and 2018 to reach over $30 trillion of investments in major developed markets. ESG-based indices have mushroomed by 14 per cent in the twelve months through June 2019. Green bond issuance reached new heights in 2019, at about $250 billion, representing close to 50 per cent increase from 2018" (UN Inter-agency Task Force on Financing for Development 2020). While nearly all elements of sustainable finance have grown, I will discuss three components here: green funds, green investments, and green bonds.

Green funds are public or private funds that finance environmental projects through grants and loans. The oldest and most common are funds for conservation, in which entities buy land in order to conserve it. For example, The Nature Conservancy, established in 1951 in the United States, has protected more than 119 million acres of land worldwide and has total net assets of nearly $9 billion.[48] Newer funds not only buy land for conservation but also support a full range of pro-environmental activities ranging from education programming to green infrastructure development. The scale of these funds is now getting quite large. The Green Climate Fund was established in 2010 by the

---

[46] World's Top Exports, "Top Solar Power and Wind Power Exports by Country," www .worldstopexports.com/best-solar-wind-exporters-powering-international-energy-sales/.

[47] This section is updated and adapted from a similarly titled section in "Effective Advocacy" (Haddad 2021).

[48] The Nature Conservancy, "Who We Are: Why We Receive High Charity Ratings," www .nature.org/en-us/about-us/who-we-are/accountability/.

194 parties to the United Nations Framework Convention on Climate Change (UNFCCC) and began disbursing funds in 2016. By 2023, it had disbursed $3 billion to support more than 200 projects, which will contribute to resilience for 666 million people and avoid emitting 2.4 billion tons of CO_2.[49] There has also been a large increase in the number and volume of funds available through national environmental funds, which are making significant contributions to biodiversity conservation worldwide.[50]

Green investing (and divesting) is a method through which individuals and institutions make investment decisions based on environmental criteria. The movement took off after 2004, when former UN Secretary-General Kofi Annan invited top CEOs to participate in an initiative that would find ways to incorporate ESG into capital markets (Kell 2018). The CEOs, banks, and many institutional investors responded, developing new "green" and "sustainable" index funds. Additionally, the "divestment" movement, in which activists around the world seek to pressure institutional investors to shift investments away from fossil fuel and other carbon-intensive businesses, helped increase the demand for more instruments that could demonstrate that their investments were not harming the planet (Ayling and Gunningham 2017).

The green finance movement gained a large boost in 2016 with the establishment of the Task Force on Climate-Related Financial Disclosures (TCFD), initiated by Michael Bloomberg and supported by the Financial Stability Board, an international body of financial regulators. Its final report and recommendations were released in late June 2017 and presented at the G20 summit in Hamburg. Financial regulators around the world began to implement the recommendations in their local jurisdictions, significantly increasing the environmental disclosure requirements for firms listed on their local stock exchanges. One can see the influence of the TCFD recommendations about ESG reporting in the frequent mention of the Task Force's recommendations in the ESG reporting guidelines of major stock exchanges.[51] As a result, there has been a rapid expansion of green investing across financial markets. By the beginning of 2020, $17 trillion in the United States was invested in sustainable,

---

[49] Green Climate Fund Portfolio dashboard, www.greenclimate.fund/projects/dashboard.

[50] The Convention on Biological Diversity, "National Environmental Funds," www.cbd.int/finan cial/0006.shtml.

[51] For example, Hong Kong: HKEX, "Consultation Paper: Review of the ESG Governance Reporting Guide and Related Listing Rules," May 2019, www.hkex.com.hk/-/media/HKEX-Market/News/Market-Consultations/2016-Present/May-2019-Review-of-ESG-Guide/Consultation-Paper/cp201905.pdf; New York: Nasdaq, "ESG Reporting Guide 2.0: A Support Resource for Companies," May 2019, www.nasdaq.com/docs/2019/11/26/2019-ESG-Reporting-Guide.pdf; and London: London Stock Exchange Group (LSEG), "Enabling Sustainable Growth," www.lseg.com/content/dam/lseg/en_us/documents/reports/lseg-sustain ability-report.pdf.

responsible, and impact investing, representing a 42 percent increase since 2018 and one-third of all assets under professional management.[52]

The area of green finance that has seen the most rapid growth has been the development of green bonds. In 2007, a Swedish pension fund issued the first green bond, seeking to reduce the risk to its investors by avoiding investments that contributed to climate change and investing in businesses and funds that promoted sustainability. Over the next few years, the World Bank, the UN, and other organizations worked to develop disclosure and investment criteria that could help ensure that green bonds actually were green.[53]

Green bond growth in Asia has been particularly large and recent. China issued its first RMB-denominated green bond in 2015, and by 2021 it had issued nearly $200 billion USD in green bonds (Deng, Xie, and Shang 2022). In 2019, Japan launched the Green Finance Network to catalyze green bond issuance in the country, and its green bond market grew by a third in 2020, to a cumulative total of $26 billion USD.[54] South Korea is now the second largest issuer of green bonds in the region after China, leading India, Japan, and Australia.[55] It also hosts the secretariat/headquarters for several of the most influential green finance organizations, including the Global Green Growth Institute and the Green Climate Fund. While Taiwan lags behind its neighbors, it too is setting new records every year. Its green bond market topped near $10 billion USD in 2022.[56]

The rapid growth in green bonds and green investing has been propelled not just by ethical considerations. Green bonds are outperforming conventional benchmarks, so they represent not just lower-risk investments but also more profitable ones (Han and Li 2022). As a result of the new reporting requirements as well as expanded financial opportunities, companies are discovering that better environmental behavior can lower their cost of capital. Whereas environmental concerns used to be relegated to companies' public relations and compliance departments, now finance departments are taking an interest, and ESG performance has become a focal point for discussions among CEOs and board

---

[52] The Forum for Sustainable and Responsible Investment (USSIS), "The US SIF Foundation's Biennial 'Trends Report' Finds That Sustainable Investing Assets Reach $17.1 Trillion" (blog), November 16, 2022, www.ussif.org/blog_home.asp?Display=155.

[53] The World Bank, "10 Years of Green Bonds: Creating the Blueprint for Sustainability across Capital Markets," March 18, 2019, www.worldbank.org/en/news/immersive-story/2019/03/18/10-years-of-green-bonds-creating-the-blueprint-for-sustainability-across-capital-markets.

[54] Climate Bonds Initiative, *Japan: Green Finance State of the Market – 2020*, March 2021 report, www.climatebonds.net/files/reports/cbi_jpn_sotm_20_02d.pdf.

[55] Climate Bonds Initiative, *Korea Climate Bond Market: Overview and Opportunities*, March 2018 report, www.climatebonds.net/resources/reports/korea-climate-bond-market-overview-and-opportunities.

[56] "Bonds Issued on Sustainable Bond Market Top NT$300 Billion," *Focus Taiwan*, May 5, https://focustaiwan.tw/business/202205050013.

of trustee members. The lower cost of capital for companies with better ESG metrics is creating a strong market incentive for above-minimum compliance and continual improvement, even among firms without much interest in the environment (Haddad 2021).

The spread of green finance, which is making more and cheaper capital available to better performing companies, is dramatically expanding incentives and opportunities for businesses to invest in pro-environmental products and practices, beyond just those firms with visionary leaders or those deploying innovative technologies. Capital market shifts affect all firms, so many companies that may not have considered their environmental impact before are now looking more closely at how they can change their operations to reduce their harm, generate more positive outcomes, and increase their access to capital.

4.4 Conclusion

Of all areas of environmental politics and policymaking, green business is perhaps where East Asia has excelled the most. Green business requires targeting investment into eco-friendly products and processes that generate profits. While it might be assumed that capitalist markets will "automatically" or "naturally" develop these business opportunities, often green business opportunities have difficulty developing and scaling up on their own. Frequently, highly profitable and exceptionally environmentally beneficial products and processes require significant upfront investment, long return-on-investment periods, and new technology, and they may be constrained by outmoded government regulations (Choi 2009; Makki et al. 2020).

East Asia's business context and its eco-developmental states were well positioned to overcome these challenges. The same coordinated policymaking systems and industrial policy tools that generated high-speed economic growth during their catch-up periods have now been directed at supporting competitive advantage in green industries (Esarey et al. 2020; Kim 2021). A long history of government collaboration with business and dense social connections have proven useful in nurturing green businesses. Modifying industrial growth tools for pro-environmental purposes, East Asian governments are using a combination of subsides, taxes, regulation, and procurement policies to incentivize the creation and expansion of eco-friendly businesses and products (Haddad 2021). In doing so, they improve their own natural environment, contribute to the health of their citizens, and support their economies by helping their companies gain competitive advantages in a rapidly expanding and highly profitable global market.

The strength of the eco-developmental state can be seen in the areas of green technology and green finance in particular. East Asian governments have made large capital investments and deployed a wide array of policy incentives to encourage and support companies in the large and highly profitable energy and automotive sectors. In both industries, companies have rapidly developed and deployed new technology that is dramatically reducing carbon emissions. As global demand for those products and services grows, those companies, their workers, and the planet all benefit. Although the finance industry has often been a laggard in terms of structural reform in the region (Katz 1998), the industry has been quick to develop and expand green finance. Throughout the region, investment firms are issuing new green bonds, banks are rewarding green companies with lower borrowing costs, and stock exchanges are making it easier for investors to identify and invest in firms with better environmental performance.

East Asia's progress in the area of green business has been led by business and has excelled because of collaboration with diverse actors including governments and NPOs. In some cases, like that of IPE in China, NPOs amplify the work of government inspectors, making it easier for financial institutions and consumers to reward environmental vanguards and punish laggards. Nongovernmental organizations can also amplify the voices of the public, helping the government see the places where policies may not be properly enforced. Foundations across the region are raising money to protect land and to accelerate green innovation.

Finally, East Asia's publics support their governments' efforts to promote green business and direct industrial planning and policy toward climate-supporting goals. Perhaps the most obvious manifestation of this has been at the ballot box, where national electorates have promoted local leaders to national office, rewarding environmental successes. As discussed in Section 3.2, Lee Myung-bak was elected president of South Korea after successfully restoring and redeveloping the Cheonggyecheon river while he was mayor of Seoul, and Chen Shui-bian was elected president of Taiwan after successfully implementing an ambitious new waste collection system as mayor of Taipei. In the national-to-local direction, Yuriko Koike was elected governor of Tokyo after serving as minister of the environment. She immediately crafted and implemented ambitious climate action plans for the city and subsequently won a landslide reelection victory in 2020.

Whereas this section on green business has offered a brief overview of perhaps the "best case" for pro-environmental policy outcomes in East Asia, the subsequent sections will show more of the failures. Section 5, on pollution, offers a more mixed story, one where tremendous progress in combating toxic pollution is tempered with pockets of persistent failure. Finally, Section 6 will show how the very same factors that have helped East Asia excel in promoting green business development – such as close government–business connections and dense social networks – are hampering the region's capacity to deal with environmental justice issues.

5 Pollution: Activists and Local Governments Work Around Central Governments and Big Business

As discussed in Section 1 of this Element, industrial pollution and its negative consequences on human health were the root causes of environmental movements across East Asia, and they continue to motivate much of the environmental activism in the region. Although the region has made tremendous progress in addressing industrial pollution, it remains a significant threat, especially to marginalized communities such as ethnic minorities, foreigners, and the poor.

Section 4 on green business highlighted ways that East Asia's eco-developmental states and its coordinated, export-oriented businesses were particularly well positioned to find pro-environmental investment opportunities that produced win-win-win-win benefits for firms, governments, society, and the environment. Green tech and green finance illustrated two investment areas where firms could make specific investments that could generate financial gains when customers and funders were able to verify their pro-environmental actions, rewarding green businesses and punishing companies that were polluting.

In economic parlance, green tech and green finance are "private goods" where some users have access, while others can be excluded, and producers can be identified. East Asia's eco-developmental states have proven adept at generating pro-business environmental policies that incentivize pro-environmental investments by businesses that then gain them greater profit and market share. This has worked well to reduce several different kinds of pollution in the region: for example, levels of sulfur, nitrogen, and carbon monoxide pollution have dropped significantly (Esarey et al. 2020).

While eco-developmental states can coordinate positive-sum solutions to private good-related pollution, they have more difficulty coping with "common pool resource" types of pollution. Common pool resources are those where the benefits of exploitation accrue to a few while the costs are dispersed among many. For example, the benefit of being able to drive a private car to a job in the city is enjoyed by the driver, but the costs of the pollution generated by the car are felt by everyone in the area. Relatedly, the costs of taking public transportation instead of driving may feel very significant to the individual making that choice, while the benefits of slightly cleaner air seem infinitesimally small (Ostrom 1990).

During the period of rapid industrialization, most of the pollution was industrial pollution concentrated near factories, a form of private good–related pollution. Governments could use their close relationship with the companies and their industrial policies to encourage firms to mitigate their polluting activities while still helping them earn a profit. Firms that consume less save money, which is positive for their bottom line. Governments can offer policy "carrots" in the form of subsidies to encourage firms to be more environmental, and once companies change from producing low-cost polluting products to high-value green products, they make more money. Thus, governments can shape the market in ways that reduce private-type industrial pollution.

In the postindustrial economy, however, most pollution is not created by a relatively small number of firms but rather by millions of individuals, through their lifestyle and consumption patterns. It is much easier to get one factory to change its production pattern than it is to convince the 20 million plus people living in a city not to drive. Furthermore, policy "carrots," like a subsidy to buy an electric car, are enjoyed by only a few, while the policy "sticks," like a gas tax, are felt by nearly everyone, making them politically unpopular.

As the example of green tech and green business illustrated, eco-developmental states do well when elite actors who are connected to multiple relevant stakeholders are able to coordinate positive-sum solutions. However, when the problem is the pollution of common pool resources, the number of stakeholders becomes much more widespread, diverse, and less elite. Furthermore, solutions often require not consuming and buying more (as in green tech or green finance) but rather buying less. Rather than win-win solutions, anti-pollution policies in postindustrial societies tend to generate a mix of winners and losers, complicating the political process.

Therefore, rather than businesses and central government officials at the forefront of pushing anti-pollution policy, it is residents, consumers, NGOs, and local governments who are the ones both creating and demanding solutions. Since they are not generally among the elites at the center of policymaking in eco-developmental states, they must find ways to generate solutions that work around the central power structures by developing grassroots solutions and cultivating allies at home and abroad. In general, activists have been the most successful when they have been able to collaborate with business and government. They have had a more difficult time gaining policy responses when solutions require significant costs and offer relatively few commercial opportunities for business.

5.1 National Efforts

Air pollution in East Asia has significantly improved across the region in the last few decades, but it continues to threaten public health and inspire political action, especially in China and in vulnerable communities in South Korea, Taiwan, and Japan. The number of people dying due to air pollution in the region has more than halved in the region, although the overall levels remain quite different. According to Our World in Data, between 1990 and 2019 the number of people per 100,000 who died due to air pollution was reduced from 280 to 106 in China, 69 to 32 in Taiwan, 75 to 28 in South Korea, and 20 to 10 in Japan.[57]

Although the situation has generally improved, air pollution is probably the most visible and politically salient environmental issue in East Asia today. Many urban residents across East Asia have an air quality app on their phone that they check daily when making decisions about whether to wear a mask, go for a walk, or take their children to a park. Public awareness and engagement around air quality issues as well as governmental policy responses are exceptionally high across the region.

5.1.1 China

Beijing won the bid to hold the 2008 Summer Olympics in 2001, and it immediately began working to address the competitors' and Olympics Committee's largest concern: air pollution. In that year, Beijing had only 12 days with clean (level 1) air, while it experienced 180 polluted (level 3–5) days, of which 23 were heavily polluted (level 4–5). The city worked rapidly to address the air pollution concerns ahead of the games. By 2008, its clean air days had increased nearly fivefold (to 62), while the number of polluted days were cut in half, to 92 (with only 8 heavily polluted days) (Chen et al. 2015). Its rapid progress was due to a whole-of-city effort to reduce pollution, starting in City Hall, and including private industry as well as the NGO sector. The city and national governments made significant environmental investments, including many that were focused on reducing the city's air pollution, such as relocating factories, shifting electricity generation away from coal toward natural gas and renewables, expanding subway and electric bus infrastructure, and engaging in widespread reforestation efforts (UNEP 2009).

The city's efforts were supported by local environmental groups. One successful effort, led by six local environmental groups in Beijing, was the 26

---

[57] See One World in Data, "Air Pollution Is One of the World's Leading Risk Factors for Death," https://ourworldindata.org/air-pollution#air-pollution-is-one-of-the-world-s-leading-risk-factors-for-death.

Degree Campaign to pressure hotels, malls, and local government buildings to set their air conditioners higher (26 degrees Celsius or above) during the summer to reduce emissions. In the summer of 2004, a group of volunteers went around Beijing with thermometers and measured the indoor temperature of a wide range of public buildings, such as hotels, malls, and government buildings. They then partnered with the media to expose how very cold (often as low as 20 degrees Celsius) these spaces were kept even during the blistering hot summer when electricity shortages were causing brownouts throughout the city.

Beijing's mayor at the time, Wang Qishan (PRC vice president from 2018 to 2023), got involved, sending his deputy mayor to hotels for surprise inspections and drawing public attention to the role that air conditioner settings play in energy conservation and the importance of energy conservation in reducing air pollution and addressing climate change. The NGO activists followed up in the fall, distributing "26 degree commitment cards" to hotels, restaurants, and malls that were hoping to attract Olympic visitors, convincing many businesses to make the pledge. Their effort received a boost the following summer when Premier Wen Jiabao declared that all government offices would set their air conditioners to 26 degrees Celsius or higher (Haddad 2021).

Concurrent with efforts to reduce smog in the capital, pollution-related protests were erupting across the country (Jing 2003; Ma 2008). Most of these were against local polluters, and the Chinese government employed numerous tactics to repress the protesters, address their concerns, and keep them from spreading elsewhere (Van Rooij 2010). Nonetheless, the level of dissatisfaction could not be ignored, reaching more than 90,000 mass incidents by the late 2000s. Furthermore, with the spread of social media, citizens had expanded their capacity to spread information, organize with one another, and pressure the government (Economy 2011).

Public awareness about air pollution was given a giant boost in 2008 when the US Embassy began tweeting about the readings from the air quality monitor on its roof. Of particular importance was the inclusion of a measure of fine particulate matter ($PM_{2.5}$), a pollutant that had not been regularly reported by the Chinese authorities but has a wide range of negative health impacts, especially on respiratory systems (Xing et al. 2016). Although the Embassy's actions initially created a political and media firestorm, the Chinese government eventually embraced public reporting of pollution data as a way of holding local governments accountable as well as touting their success in making (and often exceeding) pollution reduction targets (Bradsher 2012).

The spread of social media offered environmental activists new ways to pressure policymakers as well as providing the government with new opportunities to respond to the public. For example, on February 28, 2015, the former CCTV reporter Chai Jing posted her documentary film *Under the Dome* on an official *People's Daily* website.[58] The film was viewed more than 100 million times in the first 24 hours. It hit 200 million views by day four, when it was abruptly censored and removed from all social media in China.[59] Viewers used social networks to spread and amplify their anger, and policy-makers responded. Although it is impossible to draw a direct connection, China's 13th Five Year Plan, which was put into place almost exactly one year after *Under the Dome* was aired, had aggressive air quality targets in addition to other environmental targets that it has strengthened (Singleton and Su 2016).

Just as grassroots activists are using social media and the Internet to promote pro-environmental behavior, so has the Chinese government. In 2016 (a year after *Under the Dome*), the Ministry of Environmental Protection launched the "Black and Smelly Water Program." Through this program, citizens can use the popular WeChat app to report cases of pollution in their area (often sending pictures), and local governments are required to respond within seven business days. The reporting mechanism has also been integrated into IPE's Blue Map app to make reporting even easier (Hsu et al. 2020).

The app-based system of reporting is not perfect: in one 2020 study, only about a third of reported incidents had been fully addressed (Hsu et al. 2020). Furthermore, Iza Ding's research into enforcement, in which she embedded herself with a municipal environmental protection bureau, found that officials were more interested in appearing to be addressing the problem than actually fixing pollution (i.e., making sure a stream no longer smelled bad rather than ensuring that toxins were removed) (Ding 2020).

Although air pollution remains an intense problem in many regions in China, the situation is rapidly improving. A 2020 study that compared several key pollutants from 2014 to 2018 found that more than three-quarters of the air quality monitoring stations in the study showing reductions in CO, SO_2, PM_{10}, and $PM_{2.5}$, with some regions showing pollution levels being more than halved during the four-year period (Fan, Zhao, and Yang 2020). Through grassroots activism and the use of social media, Chinese citizens and their government are working together to pressure polluting companies to clean up while also encouraging citizens to be aware of how their own behavior, such as walking and taking public transportation or setting air conditioners higher, can contribute to the solution.

---

[58] "Chai Jing's Review: Under the Dome – Investigating China's Smog," YouTube, March 1, 2015, www.youtube.com/watch?v=T6X2uwlQGQM&t=5583s.
[59] Cui 2017.

5.1.2 Japan, South Korea, and Taiwan

Air pollution advocacy in East Asia democracies has been less individualized than that in China. As discussed in Section 2, Japan, South Korea, and Taiwan all had significant anti-pollution movements soon after their period of rapid industrialization, which resulted in significant environmental legislation and regulation, which in turn reduced the toxic levels of pollution. As a result, for the most part, none are facing the same level of pollution as China.

Japanese have the lowest engagement of people in the region on air pollution issues, largely because the air quality on the archipelago is among the best in the world and improving.[60] Although large and industrial cities continue to suffer from moderate levels of air pollution, the levels are much less than elsewhere in the region and lower than they have been in the past. In contrast to residents in urban areas in the rest of East Asia, Japanese do not generally have a daily habit of checking a pollution app before heading out for their day. While they do keep masks on hand for illness, they do not wear them as an anti-pollution measure while commuting to work, a practice commonly found elsewhere in the region.

In South Korea and Taiwan, concerns about air pollution, especially around fine dust ($PM_{2.5}$) pollution, have continued to grow in urban centers. Both places have experienced anti-pollution protests periodically in large urban centers as well as in industrial cities as residents resist expansion of polluting industrial facilities and demand government action to clean up the air.[61]

A complicating dynamic in air pollution politics in Japan, South Korea, and Taiwan is that citizens and public officials in these countries commonly blame China for their air pollution, deflecting national and local responsibility for dealing with the problem.[62] While there is considerable evidence that prevailing winds cause fine dust and other pollution to float from China to Japan, South Korea, and Taiwan (Hsu and Cheng 2019; Rodó et al. 2014; Yim et al. 2019), the issue is complicated by several factors.

First, although estimates of China's contribution to the other countries' air pollution vary wildly, none of the research suggests that China is to blame for all or even most of the air pollution in the other countries. Second, much of the industrial pollution on the Chinese coast is produced by Japanese, Taiwanese, and South Korean factories that were outsourced when labor costs rose and local

---

[60] Ministry of Environment, air pollution statistics, www.stat.go.jp/data/nenkan/70nenkan/zuhyou/y701707000.xlsx (English and Japanese); Tokyo Metropolitan Government, "Air Quality Management" summary, www.metro.tokyo.lg.jp/english/about/environmental_policy/documents/09_clean_and_comfortable_air.pdf.

[61] For example, 2017 protests in Taichung and Kaohsiung Taiwan (Everington 2017); 2017 protests in Seoul (Volodzko 2017).

[62] For example, Japan (Yokohama 2013); Taiwan (Everington 2020); South Korea (Volodzko 2017).

tolerance for pollution fell (Fuller 2008; Lin and Ma 2012; Simeon 2010). Often this outsourcing was facilitated by national governments in the form of foreign aid assistance to the host countries (Hall 2010). Finally, until 2021, when they pledged to halt financial support for overseas coal plants, South Korea and Japan were among the world's largest financiers of coal plants in China and elsewhere in Asia (Harvey 2021). Thus, although it has often been convenient for citizens and governments in Japan, South Korea, and Taiwan to "blame China" for their own air pollution problems, companies from those countries were often responsible for the toxic emissions, and their governments were often funding the construction of the power plants and industrial facilities that contributed to air pollution in the region.

Although political organizing around air pollution is not currently widespread in Japan, South Korea, or Taiwan, all three places do have a few groups that have mobilized on the issue. Ahead of the 1997 COP 3 meeting in Kyoto, where the Kyoto Protocol was signed, KikoNet was formed to connect national and global environmental NGOs together (Reimann 2003). KikoNet now has an active subgroup working on air pollution issues. Two other groups working on air quality advocacy in Japan include Toxic Watch Network and Mount Fuji Research Station.

South Korea's largest environmental organization, the Korea Federation for Environmental Movements (KFEM) is actively involved in air quality issues through its group focused on "environmental health." The Korean Society for Atmospheric Environment (KOSAE), which generally operates as a think tank and publishes an academic journal, is also active in raising awareness and promoting research about air pollution in South Korea.

Of the three democracies, Taiwan is the one with the most active and vocal anti-air pollution advocacy. Groups across the islands have regularly organized protests, especially in industrial cities like Kaohsiung.[63] Air Clean Taiwan (also called Healthy Air Action Alliance), South Taiwan Clear, and the Environmental Quality Protection Foundation all work to promote better air quality in Taiwan.

5.2 Regional Collaborations

Although air pollution is often experienced in a highly local way (e.g., the PRAISE-HK app uses street-level pollution monitoring and detailed mapping functions to help an individual navigate the lowest pollution exposure route between two places in the city), much of our air is shared.[64] As a result, although citizens might protest locally, many of the policy-relevant efforts around

[63] For the 2020 protest in Kaohsiung, see Xie (2020); for the 2017 protests in Kaohsiung and Taichung, see Everington (2017).

[64] "NASA: A Year in the Life of Earth's CO_2," YouTube, November 17, 2014, www.youtube.com/watch?v=x1SgmFa0r04.

combating air pollution are occurring on a regional and global basis, through cross-sector collaborations that involved nonprofits, governments, and businesses. In East Asia, numerous transnational agreements and working groups are seeking to address the problem. Some of the most productive areas of innovation and action have been undertaken through city networks, where municipal leaders work with one another to develop and share effective policy solutions.

5.2.1 Regional National Governmental Efforts

The national governments of East Asia are all very active in global environmental politics around improving air pollution. And yet, none of them are signatories to the most significant international agreement related to air pollution, the 1979 Convention on Long-Range Transboundary Air Pollution.[65] Although they are not signatories to the UN-based agreement, they have signed a number of bilateral and multilateral agreements related to air pollution, for example:

- 1993 Japan-Korea Agreement on Cooperation in Environmental Protection
- 1994 Japan-China Agreement on Environmental Protection
- 1996 Sino-Japan Friendship Centre for Environmental Protection
- 2015 Air Quality Joint Research Team between South Korea's National Institute of Environmental Research (NIER) and the Chinese Research Academy of Environmental Sciences (CRAES)
- 2016 Memorandum of Understanding between China's CRAES and Japan's National Institute for Environmental Studies (NIES)
- 2018 North-East Asia Clean Air Partnership

In spite of significant security tensions in the region, environmental policy has consistently been an area where regional governments have found areas of common ground, and they have forged a large number of bilateral and multilateral agreements with one another. The environmental ministers from China, Japan, and South Korea have been meeting regularly as part of the Tripartite Environment Ministers Meeting since 1999, and they have also met regularly with environmental ministers from Southeast Asia as part of the ASEAN+3 Environmental Ministers Meetings since 2002.[66]

---

[65] The text of the UN Convention, list of signatories, and ratification status can be found at https://unece.org/convention-and-its-achievements.

[66] For a fairly comprehensive overview of international cooperative activities on the environment undertaken by Japan, see the Japanese Environmental Ministry's "Bilateral/Regional Cooperation" web page, www.env.go.jp/earth/coop/coop/english/dialogue/easemm.html.

There are several multinational frameworks that aim to facilitate cooperation on environmental policy in the region. The oldest of these is the North-East Asian Subregional Programme for Environmental Cooperation (NEASPEC). Formed in 1993 in response to the UN Earth Conference in Rio, the NEASPEC member states (China, North Korea, South Korea, Japan, Mongolia, and Russia) work together to address a range of environmental issues. Transboundary air pollution is a special area of focus, generating accomplishments such as the development of the North-East Asia Clean Air Partnership, which develops a voluntary framework to address transboundary air pollution.

The Joint Research Project on Long-range Transboundary Air Pollutants in North-East Asia (LTP) is a trilateral research project among China, Japan, and South Korea. It emerged from the first North-East Asian Workshop on Long-range Transboundary Pollutants held in South Korea in 1995, and it has been led by a secretariat housed inside South Korea's National Institute of Environmental Research. The project conducts significant monitoring of a wide range of pollutants in all three countries as well as a range of impact assessments to assess the damage that air pollution causes to human health and natural environments in the region. The project has been instrumental in connecting scientific experts in the region with one another and facilitating collaborative research communities, but it has had more limited influence over policy (Kauffmann and Saffirio 2020).

The Acid Deposition Monitoring Network in East Asia (EANET) was established in 2001 as an intergovernmental initiative to address a range of air pollution issues. It has thirteen member countries from Northeast and Southeast Asia (Cambodia, China, Indonesia, Japan, South Korea, Laos, Malaysia, Mongolia, Myanmar, Philippines, Russia, Thailand, and Vietnam), and its activities are coordinated through the Asia Center for Air Pollution Research located in Japan. Its major activities focus on data collection and distribution, development and promotion of quality control activities, and technical cooperation and capacity building.

5.2.2 Transnational City Networks

One of the biggest challenges for international cooperation in Northeast Asia on any issue, environmental policy included, is the ambiguous position of Taiwan in international relations, as discussed in Section 1 of this Element. As a result of its exclusion from the UN and due to public adherence to a "one China policy" by all of the relevant governments, Taiwan is not able to take part in most international agreements. It frequently attends international meetings as an observer and participates in some meetings, but the current state of international politics hinders its ability to participate fully in global environmental politics. This is highly problematic both from the perspective of Taiwanese officials, who often feel excluded from important opportunities for collaboration, and for environmental advocates and other regional officials, who observe the high emissions in Taiwan and seek to encourage emission reductions. Transnational city-to-city collaborations have proven to be a convenient way to avoid the tangles related to national-level governmental dialogues. The mayors and environmental bureau chiefs of Taipei, Beijing, Tokyo, and Seoul can meet and collaborate in ways that the prime ministers and environmental ministers from Taiwan, China, Japan, and South Korea cannot.

Many of the most pressing air pollution issues are felt most acutely by urban residents, and municipal action is often the most effective way to address the problem. Therefore, it should not be surprising that some of the most productive transnational efforts to combat air pollution are being undertaken by networks of cities. This is true around the world and in East Asia in particular, where cities are increasingly working together in bilateral and multilateral ways to combat air pollution. They are also creating a range of city network organizations to catalyze the development and dissemination of air pollution solutions.

Perhaps the most influential of these regional organizations is Clean Air Asia, which was formed in 2001 with funding from the Asian Development Bank, the World Bank, and the United States Agency for International Development (USAID) to support city-level efforts to combat air pollution across the region. With offices in Pasig City (Philippines), Beijing (China), and New Delhi (India), the organization convenes regular conferences to bring local leaders together to share ideas, conduct original research, and offer consulting services focused especially on climate change, transportation, and low-emission development. Although it has a broader focus on climate change, CityNet also generates research and facilitates city-to-city collaborations around numerous air pollution–related issues through its program clusters focused on climate change, disaster management, infrastructure, and SDGs.

Cities across the Asia-Pacific have been working to improve regional air quality collectively. They do this through workshops and conferences as well as joint collaborative efforts that generally involve complex public–private partnerships that network cities together with private industry, local NPOs, and national funding agencies. Here are a few examples that highlight the range of collaborations taking place:

- Asia-Pacific Clean Air Partnership was launched in 2014 as a partnership between the US Environmental Protection Agency, Taiwan Environmental Protection Administration, Clean Air Asia, and the Bay Area and Coast Air

Districts. It works with regional cities, businesses, and NPOs to improve air quality in the Asia-Pacific region.

- Hai Phong Green Port City project began as a collaboration between the city of Hai Phong, Vietnam, and its Sister City, Kitakyushu, Japan. The project includes the development of new renewable energy infrastructure, electric and public transport, new municipal solid waste-to-energy, and eco restoration. Private sector partnerships include NTT, Nippon Steel, Sumikin Engineering, and the AMITA Corporation.[67]

- East Asia Clean Air Cities (EACAC) was launched in 2016 and the Northeast Asia Forum on Air Quality Improvement in Seoul, South Korea. It is sponsored by ICLEI (Local Governments for Sustainability) and the Seoul Metropolitan Government to accelerate local actions and city-to-city collaborations to improve air quality. Additional partners include the Energy Foundation China, Mongolia International University, and IGES, to name a few. The network includes ten local governments and activities range from capacity building to data collection and the development of joint projects.

5.3 Conclusion: Uneven Progress

Air pollution is one of East Asia's most pressing environmental problems – an estimated 6.5 million people die annually due to poor air quality, and approximately 70 percent of those premature deaths occur in the Asia-Pacific.[68] Public awareness has grown and spread with the development of social media apps that enable individuals to check their local pollution levels, and citizens across the region are demanding that their governments and corporations do more to combat the problem. Governments in the region have been active at the local and national levels to improve air quality. Municipal governments have been especially aggressive in addressing their local concerns as well as working in transnational networks to address the problem.

Although East Asia has made considerable progress on air pollution issues as discussed in this section, it remains an acute threat to human health and local ecosystems throughout the region. Furthermore, as with most environmental problems, air pollution is disproportionately impacting the most vulnerable populations in the region, and the variation can be enormous. In one frequently cited study, the 500 million residents of northern China had a reduction of life

---

[67] OECD (2016); City of Kitakyushu and City of Hai Phong, "Green Growth Promotion Plan of the City of Hai Phong," May 2015, https://esci-ksp.org/wp/wp-content/uploads/gravity_forms/2-78aae73684240255535d280ca98acba9/2018/10/Green-Growth-Promotion-Plan-of-Hai-Phong-City.pdf.

[68] UNEP, "Restoring Clean Air," www.unep.org/regions/asia-and-pacific/regional-initiatives/restoring-clean-air.

expectancy of 5.5 years on average (or a total loss of 2.5 billion life years) compared to counterparts living in the southern part of the country (Chen et al. 2013). South Korean communities with a high proportion of minorities suffered higher levels of toxic emissions than communities with fewer minorities (Yoon, Kang, and Park 2017), and Indigenous communities in Taiwan are suffering from pollution-related chronic obstructive pulmonary disease at higher rates than the rest of the population (Chan et al. 2014). These and other environmental justice issues will be discussed more extensively in the next section.

Thus, although East Asia has made considerable progress in combating the toxic legacy of its rapid industrialization and high-growth era, that progress has not been uniform. Citizens across the region continue to pressure their governments to tighten regulations, punish polluters, and clean up their air, water, and soil. In many ways, they have been successful. Working at the grassroots level and nationally and transnationally through a wide array of organizations, activists have succeeded in forcing governments and businesses to clean up.

Activists have been most successful when they have been able to partner with government and business to address the problems. They have had less success when they operate in direct opposition to governments, and political marginalized communities such as migrants and Indigenous groups have had the hardest time addressing pollution in their communities. Therefore, pollution is an environmental issue area where East Asia's outcomes have been significantly more mixed than in green technology and finance, where it has been more successful.

6 Environmental Justice: Marginalized Communities Still Losing Out

If green business is the area of environmental policy where East Asian countries have been making the most forward progress, and pollution is an area where outcomes have been mixed, environmental justice is where we see East Asian countries struggle the most. As discussed in Section 1, the developmental state model prioritized government–business collaborations that promoted economic growth, and it was spectacularly successful in achieving that growth. When environmental pollution caused by rapid industrialization began threatening lives and livelihoods, the people pushed back and demanded that their governments and their corporations respond. The result was a rapid incorporation of environmental concerns into the developmental state model, which catalyzed green business and made significant strides in addressing health-threating pollution.

However, progress was not universal, nor was it even. East Asian governments have made significant strides in addressing health-threatening pollution in places and among people with a political voice. When business and central government elites can work together to create positive-sum solutions, as they have in green business development, progress has been rapid and dramatic. When powerful elites central to East Asia's eco-developmental states cannot find easy profits, progress lags. In the area of air pollution, local activists, NGOs, and local governments have found ways to work around obstructionist actors in the central government and cultivate allies, crafting a few solutions, although they have tended to be more ad hoc, piecemeal, and less dramatic than those in the green business space. Indigenous, poor, and minority communities are even farther from the centers of power, so they have often borne the highest levels of pollution and are usually the last to receive relief.

This section takes a closer look at environmental injustice in East Asia in three parts. First it will examine environmental justice issues facing Indigenous communities, which tend to be rural and geographically far from the capital cities. Next it will explore the environmental problems plaguing minority and immigrant populations, which can be rural but are more often found on the margins of the largest cities and in second- and third-tier cities. Finally, it will show some ways that East Asia is exporting environmental injustice by moving its polluting practices not only to its own vulnerable communities but also to marginalized communities abroad. After discussing these three forms of environmental injustice, this section will explore one area where there has been some progress being made through ecotourism, which offers a way for the eco-developmental states to channel their pro-business efforts in ways that can benefit Indigenous communities.

6.1 Indigenous Communities as Targets

Perhaps the greatest barrier to environmental justice in East Asia is the myth of homogeneity that is common across the region. On the one hand, the myth that "everyone" is "the same" is not entirely unreasonable, since China, Japan, South Korea, and Taiwan are more ethnically homogeneous than many places around the world, with more than 90 percent of their populations belonging to the same ethnicity. On the other hand, images of a "harmonious society" where everyone looks the same, talks the same, and shares the same cultural values are the direct result of political choices that discourage immigration and emphasize assimilation. Although many of the places are now moving toward a more multicultural understanding of themselves, the myth of cultural homogeneity persists across the region (Chung 2020; Leonard and Lehmann 2019). Popular ideas of homogeneity make it particularly difficult for citizens and governments in the region to identify and address environmental justice issues, especially when they are related to Indigenous and migrant communities who are "different" from the majority.

The problem of environmental harm to Indigenous communities is particularly acute and visible in Taiwan, which has the greatest ethnic diversity and the largest Indigenous population in East Asia. Indigenous populations in Taiwan have experienced numerous waves of displacement from their lands and have been targeted for polluting industrial facilities and unwanted coal plants (Chi 2001) as well as nuclear waste facilities (Chang 2020; Fan 2006). As is the case everywhere, the issues are complex. To choose just one example, although Taiwan's Indigenous communities have been targeted as locations for nuclear waste dumps, and they currently house 80 percent of the island's nuclear waste (Huang, Gray, and Bell 2013), not all communities are equally opposed. Taiwan's main waste facility is located on Lanyu Island, and the Indigenous Tao people have been engaged in a decades-long fight against the "temporary" storage of nuclear waste on their lands. In contrast, in 2008 and 2011 Nantian, a Paiwan village, was selected as a potential site for the nuclear waste, and those villagers expressed much less opposition to hosting the waste, especially once the government promised that economic investment would accompany the new waste facility (Chang 2020).

For many decades, environmental justice for Indigenous peoples did not garner much national recognition in Taiwan, but this has been changing recently. Celebrating Indigenous identity and expanding Indigenous rights have become increasingly important national political issues as the political and social rifts between "Taiwanese" and "mainland Chinese" identities have grown (Cabestan 2017; Fuyan 2019). For the first time, in 2016, Taiwan's current president, Tsai Ing-wen, apologized for the atrocities committed against Indigenous people, and during her 2020 re-election campaign (which her party won by a landslide) she offered to financially compensate families in Lanyu for housing the nuclear waste. The mayor of Lanyu rejected the compensation, indicating that, while he appreciated the offer, his people would maintain their fight to get the waste removed (Aspinall 2019).

Environmental justice issues in China have somewhat similar dynamics to the ones in Taiwan, since most of the battles involve the central government and state-owned energy companies targeting minority ethnic groups and their sacred lands for resource-related development projects. However, in contrast to democratic Taiwan, where these conflicts have entered the national political spotlight and are a focal point for interparty contestation, information about identity-based conflicts in China is more hidden, since they are politically problematic for the CCP. Officially, Xi Jinping and the CCP frequently point to China's "successful" incorporation of ethnic minorities into the "Chinese dream" as one more piece of evidence for the superiority of the "Chinese model," in contrast to the drama and violence of identity politics that often plague democracies (Xin 2018: chap. 1). Therefore, any evidence that China is experiencing ethnic conflict is ruthlessly repressed (Rozenas 2020).

Here are two examples of the kinds of environmental justice conflicts that are occurring in China: In Tongpo a group of Mongol pastoral people sought to protect a sacred mountain and conducted multiple protests in 2005 and 2006 against mining activity and the construction of a coking plant at the mountain's base. They failed in their effort – the mining continues and the coking plant was built (Chuluu 2021). In Panguanying, a rural Hebei village, residents opposed the construction of a waste incineration plant. In 2009, they created a rural–urban environmental network that eventually challenged the government's environmental review process in court. They won a halt to construction in 2011, although the project has been paused rather than canceled (Johnson, Lora-Wainwright, and Lu 2018).

The Chinese community facing the most intense environmental injustice at the moment is the Uyghurs in Xinjiang. The autonomous region was the site of China's nuclear testing in the 1960s into the 1990s and is home to much of China's coal mining (and burning). Heavy industrial development, intensive cotton farming, and climate change–accelerated desertification have combined to reduce and contaminate the region's water supply (Scull 2008). Not surprisingly, these intense environmental pressures have contributed to the political and social unrest in the region (Baranovitch 2019).

Japan is highly urban with very small Indigenous populations, and yet environmental injustice targeting Indigenous peoples occurs there as well. In one extraordinary case, official recognition of Japan's Indigenous Ainu people as a distinct ethnic group was a direct result of a legal finding of environmental injustice. In the 1980s, as part of Japan's extensive construction boom (Woodall 1996), officials began the process of negotiating land purchases in order to construct the Nibutani Dam in Hokkaido. Two of the property owners were Ainu and refused to sell. Their land was seized, and construction was started in 1986. In 1993, a lawsuit was filed calling for the revocation of the expropriation decision, claiming the government had ignored the importance of the sacred land to the Ainu when making the decision. In 1997, in a landmark decision, the Sapporo district court concluded that the Ainu people had the right to enjoy their cultural land, that they fit the definition of Indigenous people, and the Japanese government had neglected to take this into account when building the dam. Since the dam had already been built and the damage done, the ruling was unable to restore the land to its owners. Although the plaintiffs lost the battle, they won a greater war: the 1997 verdict was the foundation for the 2008 unanimous decision of the Japanese Diet to recognize the Ainu as an Indigenous people in northern Japan (Maruyama 2012).

6.2 Marginalized Minorities in the Cities

While the injustice faced by Indigenous groups in East Asia has been acute, environmental justice problems in urban communities are affecting more people, tens of millions across the region. As discussed in Section 5, air pollution continues to trouble much of East Asia, and marginalized communities – especially ethnic minorities, immigrants, and the poor – bear a disproportionate burden. They also suffer from other urban environmental harms such as poor waste management, polluted water, lack of green space, and so on.

In some places, such as China, the research has been extensive enough to document with gruesome precision the health cost to those living in these targeted communities. In one study of the "model city" of Dalian, suburban residents suffered 50 percent higher rates of cancer than urban residents of the same city because factories, power plants, and municipal waste facilities had been relocated to the suburbs, and wealthier farmers utilized more polluting fertilizers and pesticides (Zhang and Liu 2021). Another study of China's "garden city" of Hangzhou demonstrated how a committed public effort succeeded in greatly expanding the green space in the city (about 40 percent of the area in Hangzhou is now "green") (Wolch, Byrne, and Newell 2014) but also exacerbated environmental harm to the city's poor and vulnerable populations. For example, when parks were constructed adjacent to highways, residents were exposed to higher levels of harmful air pollution. Similarly, when parks were designed primarily for aesthetic purposes, they did not include opportunities for active recreation and had limited outdoor play spaces for young people. As a result, those urban communities were denied the social and health benefits they should have received from their new green spaces (Wolch, Byrne, and Newell 2014).

Echoing those studying China, scholars investigating environmental justice in South Korea have found similar patterns of inequality in South Korean cities. One study examined the toxic release inventory levels in all 230 local governments in South Korea and found that jurisdictions with higher percentages of minorities, greater industrial land use, and less political activity suffered greater levels of pollution, even as economic status did not affect the siting of hazardous facilities (Yoon, Kang, and Park 2017). Another study of Seoul's parks found that neighborhoods with more minority residents in the city had significantly lower access to public parks and green space than neighborhoods with fewer minority and immigrant residents (Oh, Kim, and Sohn 2020).

Although environmental equity issues persist in Japan (Yasumoto, Nakaya, and Jones 2020), the country also offers one of the most hopeful stories of how these kinds of conflicts can be resolved. Tokyo's now-famous "Garbage Wars" were fought in the early 1970s when the residents of the (relatively rich) Suginami ward refused to build a new waste incinerator plant to handle their growing volume of garbage, preferring to continue to truck it to the neighboring (and comparatively poorer) Koto ward. In 1971, residents in the Koto ward were fed up and forcibly blocked trucks from Suginami from entering Landfill Number Fifteen. After several more protests and years of negotiation, Suginami finally agreed to build an incineration facility of its own to handle the waste produced by its residents. Furthermore, rather than a smelly landfill, the new facility show-cased high-tech designs to minimize pollution as well as new co-generation technology that used the heat from the incinerator to warm a nearby swimming pool and senior citizens' home. New community spaces and a library were housed in the same complex. Over time, residents in Tokyo, and across Japan, came to recognize that every family and every community should take responsibility for their own waste, reducing its volume, increasing reuse and recycling, and finding environmentally and socially just ways for final disposal (Siniawer 2018).

6.3 Exporting Environmental Injustice

As discussed in Section 1 of this Element, East Asia's developmental states built their economies using export-oriented growth. As the process of rapid industrialization and economic growth evolved, their companies became globally competitive. As market share grew, citizens became more demanding, requiring higher wages and higher environmental standards. As a result, East Asia's global corporations, like their counterparts in North America and Europe, began to diversify their supply chains, moving manufacturing facilities to places with lower labor and environmental standards. One of the consequences of these supply chain shifts is that East Asian countries exported some of their environmental injustice to their neighbors.

The exporting of environmental injustice began with Japan, which began offshoring its manufacturing first to the other countries of Northeast Asia (e.g., South Korea, Taiwan, and China) and then to Southeast Asia (e.g., Vietnam, Malaysia, Philippines, and Indonesia) as well as Latin America. Japan's actions were soon followed by those of its neighbors, which also began moving manufacturing abroad to Southeast Asia, Latin America, and Africa. While the new factories did contribute to economic growth in host countries, they also compounded environmental injustices as marginalized communities were exploited for economic gain.

Furthermore, the structure of the developmental state meant that corporate efforts to extract economic gains were facilitated by their governments, who often paved the way for private investment with government-funded infrastructure distributed through various forms of foreign aid. In contrast to the corporate investment, which was always justified as a business decision that anticipated a return on investment, government aid was often justified on moral grounds as a way for the now-rich countries of East Asia to offer their technology and financial support to less developed countries (Dal et al. 2021; Hall 2010; Nedopil 2021).

China is the largest of the East Asian economies and the one most frequently accused of exporting environmental injustice. It has long maintained that its commitment to political "non-interference" in the recipient country is a morally and politically better way to deliver foreign assistance than the methods practiced by the member countries of the Organisation for Economic Co-operation and Development (OECD), the World Bank, the International Monetary Fund (IMF), and similar Western-country-dominated entities that commonly added political and governance conditions to their foreign aid (Lengauer 2011). Partly as a result of its philosophy about the appropriate role of the donor country, accusations of environmental injustice have continually plagued Chinese foreign investment. Whether the investment projects were dams (Siciliano et al. 2019), mines (Shapiro 2019), or timber (Ekman, Wenbin, and Langa 2013), critics have argued that Chinese investment in developing countries has prioritized Chinese economic gains over social and environmental improvement in aid-recipient countries.

Recently, China's ambitious Belt and Road Initiative (BRI), which is set to make $1.2–$1.3 trillion USD in infrastructure investments in more than a hundred countries around the world, has dramatically increased the scale of Chinese investment abroad as well as the accusations of environmental injustice (Chatzky and McBride 2020). Partly in response to critics, in 2019 President Xi announced that environmental considerations must be integral to a "Green BRI," which intends to build "high-quality, sustainable, risk-resistant, reasonably priced, and inclusive infrastructure" (Goh and Cadell 2019). While there has been some skepticism about China's willingness to uphold this commitment (Coenen et al. 2021; Zhou et al. 2018), there is also evidence that through its commitment to green business China is contributing to greener investments and paying greater attention to environmental justice issues than before (Liu and Xin 2019; Zhang et al. 2021).

Accusations of exporting environmental injustice and contributing to climate change have also been directed at Japanese and South Korean ODA. Like China, Japan and South Korea have actively engaged in foreign development assistance across a wide range of projects. They too have used "Green ODA" as a way of gaining a kind of "triple return" on their investments. Green aid projects help developing countries deploy green technology faster, help support donor countries' green tech industry, and contribute to positive environmental outcomes (Liu 2016; Hall 2010; Han 2015). Unfortunately, also like China, until very recently Japan and South Korea have also generously financed coal infrastructure abroad. While both countries have argued that their "clean coal" technology was better than the "dirty coal" that had been planned by the host countries, there is no doubt that the new coal plants negatively contribute to climate change and inflict additional pollution on vulnerable populations. In 2021, both South Korea and Japan agreed to halt funding for coal plants abroad as part of their Net Zero by 2050 commitments (Liu, Wang, and Wang 2021).

Finally, in addition to ODA, the problematic actions of East Asian companies when they expand abroad have sparked public protests and accusations of environmental racism in a number of different countries around the world. For example, the Waorani people in Ecuador have filed a lawsuit against Chinese-owned PetroOriental, accusing it of contaminating their ancestral lands (Sanchez 2020). The Black community in St. James, Louisiana, is fighting the construction of a new plastics complex by Taiwan's Formosa Plastics Group, accusing the company of environmental racism (Mufson 2021). The South Korean palm oil giant Korindo is accused of setting fires that are deforesting the traditional lands of the Mandobo tribe in Indonesia (Amindoni and Henschke 2020). Japanese environmental activists have spent decades using what Simon Avenell calls Japan's "environmental injustice paradigm" to fight the environmental harm caused by their own companies at home and abroad (Avenell 2017).

6.4 Ecotourism Offers a Ray of Hope

One area where the pro-business strengths of East Asia's eco-developmental state have been able to benefit Indigenous and marginalized communities is through the promotion of ecotourism. In the last few decades, as East Asia's economies shifted toward a service-based economy, tourism has boomed. Interestingly, in spite of high diplomatic tensions among the countries, regional tourism also boomed. For example, Chinese tourists traveling to Japan increased nearly tenfold from 1.4 million to 9.6 million between 2010 and 2019. Over the same period, Taiwanese tourists jumped from 1.2 million to 4.9 million, and South Koreans from 2.4 million to 5.6 million. By 2019, the volume of inbound tourists to China had risen to 145 million people (75 percent from Asia), and international tourism was dwarfed by the volume of domestic tourism, which reached more than 6 billion people during the same year. The tourism industry in China accounted for 11 percent of its total gross national product (GNP) in 2019 (Xinhua 2020). South Korean and Taiwanese tourism enjoyed similar growth – foreign arrivals to South Korea jumped from 8.8 million to 17.5 million between 2010 and 2019,[69] and those to Taiwan rose from 6 million to 12 million over the same period (Steger 2020). The COVID-19 pandemic put a halt on all tourism in the region for several years, although it does appear to be bouncing back (Pitrelli 2022).

Globally, sustainable Indigenous tourism has become a popular way to promote the economic development of Indigenous communities while empowering them to preserve and enhance their cultural heritage as well as protect the environment (Zeppel 2007). Although sustainable Indigenous tourism often has to balance the positives of economic development for Indigenous communities against the potential for commercialization and environmental exploitation if the number of visitors grows too large, the approach is often seen as a win-win-win opportunity for governments to support economic and social development of Indigenous communities while also addressing environmental concerns (Carr, Ruhanen, and Whitford 2016; Zhuang et al. 2017). Furthermore, ecotourism as an industry is quite flexible, able to accommodate a wide variety of different local environments as well as cultural values. One review of the ecotourism literature in South Korea found the emergence of a "South Korean model" of ecotourism that reflected the prevalence of "nature-based" attractions influenced by South Korea's Confucian, Zen, and Taoist philosophical heritage, a strong emphasis on "education," Confucian values of self-cultivation, and "sustainability" interpreted as compatible with Confucian ideas of harmony (Lee, Lawton, and Weaver 2013).

All of the East Asian places studied here have found some success in utilizing ecotourism as a method to address environmental justice in their countries. I will offer one example from each place to illustrate some of the complexities involved in ecotourism as an effort to address environmental injustice. The first case is Indigenous River Closure, a mountain river ecotourism movement in Taiwan in which Indigenous tribes use ecotourism as a means of economic development and disaster resilience and recovery. In one study of seven villages containing three different ethnic groups (Bunun, Paiwan, and Rukai) along the Kaoping River, the ecotourism projects helped bring young villagers back from the cities, leading to community revival, cultural preservation, and enhanced tribal identity. They were also able to build ecological resilience, spreading

---

[69] World Data Atlas, "Republic of Korea – Number of Arrivals," https://knoema.com/atlas/Republic-of-Korea/Number-of-arrivals.

ideas about river conservation and teaching others that "a fish's life is more meaningful and valuable in the river than in someone's belly" (Shie 2020). However, paradoxically (and disappointingly), national governmental and NGO enthusiasm for scaling up successful projects often backfired – outside support would bring in needed financial resources but also "reduced local spontaneity and unintentionally manipulated villagers to pursue goals favored by outsiders" and led to factional infighting (Shie 2020).

Of the seven villages studied, two had good outcomes, four had medium outcomes, and one had a poor outcome. Given the overall positive results and the fact that the River Closure groups "improved solidarity and environmental consciousness within communities, built community connections, solved collective environmental issues, raised environmental awareness, enhanced public participation and regained ownership of their environment" and some "generated economic rewards," the Pingtung County Government decided to promote River Closure to more than eighty rivers, enabling the Indigenous groups to compete with powerful sports fishing associations for policy influence (Shie 2020).

An extensive study about an ecotourism development plan in Jiuzhaigou National Nature Reserve in China that involved public–private collaboration with Tibetan villagers had findings similar to those of the River Closure movement in Taiwan: ecotourism projects generated economic growth and could enhance local ethnic identity. However, if decision-making structures were top-down rather than participatory, the projects that were intended to raise up the local community could instead contribute to inequity, loss of cultural identity, and ecological degradation (Yang 2019). Additionally, lack of meaningful power sharing led to a large gap between the perceptions of the local Tibetans and those of the professional staff on the reserve. For example, the Tibetan villagers thought that the economic benefits of the Reserve were not equally shared, they were relegated to low-level jobs, and infrastructure improvements were designed to benefit tourists, not locals. In contrast, staff members looked at the same collaboration and thought that the programs had generated remarkable income growth, the villagers were the richest Tibetans in China, and the area now boasted an airport, highway, drinking water pipelines, and greater electrification, which benefited the whole community (Yang 2019).

One promising area of ecotourism that is being developed in South Korea (and around the world)[70] is eco-organic farm tourism. In this form of ecotourism, visitors experience traditional rural hospitality and can enjoy an authentic

---

[70] Worldwide Opportunities on Organic Farms (WWOOF; wwoof.net) has been sending visitors to organic farmers for cultural and educational exchange for fifty years.

experience while staying, learning, and sometimes working on an organic farm. One study of thirty-eight farms offering eco-organic tourist experiences found that this form of ecotourism was able to avoid some of the problems faced by ecotourism. In particular, community engagement in decision-making was guaranteed, since the experiences were developed and run by the farmers themselves. Additionally, the farmers were the primary beneficiaries of tourist spending. They made an effort to source additional materials and services locally, so they were able to ensure that local community members were the primary beneficiaries of the added income generated by the tourist activities (Choo and Jamal 2009).

Furthermore, although the famers did develop activities to accommodate tourists' interests, such as work-study programs, folk art, cooking classes, and field trips, they had a rich cultural heritage on which to draw. "Nature" and "culture" were not viewed as separate; cultural activities were interconnected with ecological activities. "All respondents expressed a common belief that they had a sacred duty to preserve the land since they not only inherited it from their ancestors but also borrowed it from the next generation" (Choo and Jamal 2009). As in the Taiwanese and Chinese cases, South Korean farmers expressed a tension with respect to scale – if the number of tourists grows too high, it degraded the environment and the experience. In some ways, South Korea's eco-organic farm tourism was able to address some of the scale-related problems because each operation had to manage within the bounds of its own farm. However, those limits on scale also limited the added income possible from the tourists, and farmers remained concerned about their financial futures (Choo and Jamal 2009).

As a final example, in 2005 one of the most spectacular parts of Japan, which lay within the traditional lands of the Ainu people, Shiretoko National Park, became Japan's third UNESCO World Natural Heritage site. The designation was part of Japan's multi-decade effort to expand its tourism industry (Lehney 2003). As with the case of Jiuzhaigou Nature Reserve discussed earlier in this section, the establishment of a national government-supported and much-visited park inside the historical territory of an Indigenous people has generated a local ecotourism boom in the area with complex effects on the Indigenous community.

At the time of the UNESCO application, the Japanese government denied the presence of the Ainu people and did not recognize them as an Indigenous community. However, even after their recognition as an official Indigenous group in 2008, the Ainu have been sidelined in the governance structure of the park (Lewallen 2016). The park's current website barely recognizes the land's relationship to Ainu people and culture. Although the establishment of the park and its status as a UNESCO site can be seen as a contemporary instance of settler colonialism, where the Japanese government is (again) asserting the right to control "its" land, it has also created new opportunities for the Ainu to rejuvenate their culture and broaden their networks of support (Lewallen 2016). The Ainu-led ecotours "offer the possibility of rejuvenation: transforming Ainu relations with the land toward renewed stewardship and reciprocity, of being looked after by the land, rather than merely caring for the land" (Lewallen 2016: 8).

Thus far Ainu ecotourism has not generated great economic gains, but it has created opportunities for the Ainu to reenergize their community and gain greater external recognition. Through the ecotours, urban Ainu reconnect with the traditions of their ancestors and non-Ainu gain exposure to Ainu worldviews. The park, as well as the Ainu Indigenous ecotourism in and around it, "provides a context for a host of practices to resignify Ainu" (Lewallen 2016: 1). With new opportunities for storytelling, passing traditional practices on to new generations, and educating outsiders about the value of a worldview that "cultivates awareness of the interconnectedness of all living beings" (Lewallen 2016: 11), ecotourism is making an "Ainu contribution for Japan toward advancing sustainability and balance with the nonhuman world" (Lewallen 2016: 11).

6.5 Conclusion: Environmental Justice in East Asia

As this section has demonstrated, although East Asia has made rapid and remarkable progress in improving its environment in many areas, it has struggled with environmental justice. Cultural myths of homogeneity have made it more difficult to recognize and address environmental injustice when it happens. As a result, Indigenous peoples, minority groups, and other marginalized communities continue to suffer disproportionately from environmental pollution and face higher risks associated with climate change–induced disasters and displacement. Indigenous villages in rural areas as well as poor and minority communities in cities have been targeted for pollution in all the countries of East Asia. Corporate and governmental actors in the region have then compounded their damage by exporting environmental injustice abroad, moving polluting facilities away from their own countries and locating them in vulnerable communities abroad.

As local groups organize and public awareness grows, some incremental progress is being made. One particularly hopeful opportunity can be found in sustainable Indigenous tourism, which can, if organized correctly, offer a chance for Indigenous communities to bring economic development to their areas while also protecting, and sometimes even enhancing, their ecological environment. Unfortunately, this small ray of hope only underscores how difficult it has been for the eco-developmental states to address environmental justice. When victims and advocates are disconnected from decision-makers, it is extremely difficult for them to influence policy and achieve the positive environmental solutions that they both need and deserve.

Another small sign of hope for environmental justice has emerged in the aftermath of the COVID-19 pandemic and the global Black Lives Matter movement. The pandemic shone a giant spotlight on the connection between environmental, health, and social justice as people in marginalized communities around the world suffered disproportionately. Whereas the 2008–9 economic crash was seen as a financial crisis with a financial solution, the pandemic was seen as a triple crisis – health, economic, and political (Tiberghien 2021) – that would require a comprehensive set of policy responses. In 2009, East Asian governments targeted green businesses as a way to improve global competitive advantage while also improving their environments. In 2021 and 2022, regional governments increasingly focused their policies and their finances on those areas that could benefit the economy, the environment, and society, especially in vulnerable communities. In some cases, they even talked about fundamentally reshaping economic systems in ways that would prioritize social well-being, seeking to ensure that the most vulnerable members of their communities also benefit from economic growth. This shift, or hint of a shift, will be discussed at greater length in the final section.

7 Conclusion

The most important takeaway from this Element is the following: If environmental progress is possible in East Asia, it should be possible almost anywhere. A country does not need to be rich or have strong green parties, a robust democratic system of government, or a well-funded NGO sector to make significant environmental progress. It is possible for citizens in very diverse political contexts to pressure their local and national officials to make positive change. Small, grassroots organizations can connect with one another to mobilize consumers to reward companies that act in environmentally responsible ways and punish companies that do not. Financial institutions and businesses can be positive agents of change, offering funding to promising, innovative companies that are creating climate solutions. Governments, banks, and financial markets can turn away from companies that pollute, waste, and are environmentally irresponsible. Economic wealth and democratic elections are not necessary for any of these community-based, consumer, financial, and bureaucratic actions.

Environmental advocates across East Asia are operating under very difficult and very different political conditions. They did not have a single strategy that enabled them to be effective, but rather they drew on their personal and professional connections, their local traditional cultural resources, entrepreneurial creativity, and other sources to find innovative ways to pressure and inspire political and business leaders to include environmental considerations into their policymaking.

Sometimes, they did this through lawsuits and public protests that grabbed the public's attention, shamed political leaders, and financially threatened businesses. More often, however, their success came by finding win-win-win-win areas of collaboration that benefited citizens, governments, companies, and the planet. Across an extraordinarily diverse set of circumstances – from Beijing neighborhoods threatened by smog, to Indigenous lands in the mountains of Taiwan facing intense floods, to the boardrooms of South Korean auto companies and the garbage collectors and housewives in Tokyo – those seeking to improve the quality of life in their own neighborhoods worked with governmental policymakers to find creative solutions that simultaneously improved their own household's circumstances while also improving the livelihoods of those around them and contributing to a greener planet.

The general pattern of East Asia's eco-developmental states has been one where the most environmental progress is made when positive-sum solutions can be crafted by elite government and business allies working together. As the affected people and advocates get farther from the centers of power, it becomes more difficult to create and implement pro-environmental policies. To the extent that citizens, NGOs, activists, Indigenous communities, and so on can cultivate allies in elite policy circles, they are often able to make significant progress. When they are not able to access the centers of power, it is more difficult for them to make big, lasting changes. However, East Asia demonstrates that even when people are far from the center of power, they are not powerless. They can develop local solutions to solve the problems in their own communities. They can work around central governments and big business, networking with allies to spread their creative and effective solutions to other communities at home and abroad.

As this Element demonstrates, pro-business, pro-government environmental solutions are not always possible. They are particularly difficult when the people facing environmental harm come from politically marginalized groups or when mitigation requires costly investments and generates low profits. However, East Asia demonstrates that win-win-win-win options may be more available than previously thought. Even undemocratic governments not subject to electoral threats can find political benefits in greener policies. Even polluting energy companies and giant financial institutions that profited from fossil fuels can find ways to make money from climate-friendly enterprises.

East Asia's pro-business path toward better environmental policy underscores the challenges related to addressing environmental justice. Environmental injustice is a problem faced by all regions of the world (Adeola 2000; Fuentes-George 2016; Robinson 2018; Schlosberg 2009), and the East Asian experience demonstrates that this issue area is particularly resistant to pro-business solutions. Indigenous peoples, immigrants, and ethnic minorities are politically marginalized around the world, and they have difficulty convincing political leaders to prioritize their concerns. Their challenges are even greater in nondemocratic contexts where the lack of a free press often means that their voices are silenced, their stories are censored, and their bodies are imprisoned or worse (Poulos and Haddad 2016).

Climate change is a global problem, and natural environments and ecosystems are connected, so solutions must be local, regional, national, and global. Governments in the region and around the world are increasingly recognizing that there cannot be a piecemeal solution to climate change. Only a system-wide adjustment toward a greener and low-carbon society will allow some semblance of our current social and economic systems to continue. Furthermore, policies focused on improving the environment can both mitigate the worst damage related to industrial pollution and climate change and generate political dividends for leaders who are successful in improving environmental outcomes.

East Asia's eco-developmental states have focused government resources and policy toward supporting the green businesses and services that can enhance the quality of life for citizens at home as well as the national competitive advantage for companies operating abroad. They include an increasingly diverse range of perspectives in their policymaking to maximize the efficiency and efficacy of their resource allocations and policy choices. These practices have delivered a high standard of living to their populations while also cleaning the air, decontaminating the water, and expanding public green space. In return, although particular leaders may be shuffled in and out of power, the public trusts and supports its government.

7.1 Is China Still an Eco-developmental State?

I began writing this Element in late 2019 before there were even whispers of the health crisis that would eventually kill more than 7 million people around the world. Through both the shutdown and the slow reopening, East Asian governments responded as eco-developmental states would be expected to respond – they collaborated with businesses and civil society actors to direct large government funding toward industries and services that served to enhance national competitive advantage, improve the economic well-being of citizens, and address environmental concerns.

In January 2022, it appeared that East Asian governments had all reinforced all three pillars of their eco-developmental statehood: they focused on enhancing national competitive advantage in green business, consulted with diverse stakeholders when making policy, and enjoyed strong public support for their policies. However, a year later, conditions had changed in China such that it is no longer clear whether it should "count" as an eco-developmental state. As indicated in Section 3, there have long been scholarly debates about whether it has ever been appropriate to classify China as a developmental state, let alone an eco-developmental state. Over the course of 2022, while the other states in East Asia reinforced their environmental policies, expanded the diversity of stakeholders included in policymaking, and actively sought to retain and improve their public support, China appeared to go in the opposite direction in all three dimensions.

Of the three pillars of eco-developmental statehood, China remains quite strong in the first, namely supporting green technology as a means of strengthening national competitive advantage. It has continued or even expanded some areas of state support for clean energy infrastructure and green technology as a component of its pandemic recovery plan. However, many of its renewable energy projects, both those at home and those it supports abroad, have been put on hold due to funding and supply chain restrictions (Bashir et al. 2022; Tian et al. 2022).

The second pillar of eco-developmental states is the inclusion of diverse stakeholder perspectives in policymaking, and China seems to be retreating from this pillar. The only woman in the country's top policymaking body, the Politburo, stepped down in late 2022 and has not been replaced. The twelve-person China Council for International Cooperation on Environment and Development has only one woman (who is from Europe) and only two members with significant NGO experience. Many NGOs in China have closed their doors, and those that remain are finding it more difficult to access policymakers (Holbig and Lang 2022). Thus, it appears that the range of perspectives included in policymaking in China is narrowing rather than diversifying (Fang and Lai 2022; Xia 2022).

The final defining feature of eco-developmental states is broad public support for governmental policy. As discussed in Section 3.4, the "blank sheet" protests of December 2022 indicated widespread citizen dissatisfaction with the government. In response, the government abruptly ended its zero-COVID policies and detained any remaining protesters. By early 2023, there were no longer widespread street protests in China.

In sum, China's commitment to green technology and services as a priority for its industrial policy, which is intended to build global competitive advantage, appears to remain strong (Lin and Zhou 2022; Sun et al. 2022). However, recent political purges have resulted in a largely homogeneous group of people making policy, and widespread unrest at the end of 2022 suggests that the government may no longer have the support of its people. Both factors might be merely temporary setbacks associated with the difficult adjustment to a post-pandemic situation, although it is impossible to know at this moment.

7.2 East Asia's Vision for a New Sustainable, Inclusive, Economy

This Element has argued that East Asia's environmental politics have been shaped by the evolution of its developmental states into eco-developmental states. As East Asia's growth-at-any-cost developmental states achieved economic prosperity and polluted their environment, they began incorporating environmental priorities into their policymaking. As they shifted from "catch-up" economies to global leaders, they diversified the perspectives included in policymaking. Their eco-developmental state structure enabled them to make great strides in green business and finance while addressing many pollution concerns. At the same time, they continue to struggle with environmental justice when victims were far from the center of power and solutions required significant financial investments without the possibility of generating profits.

Like the pollution crises of a generation earlier, the COVID pandemic shone a giant spotlight on the close connection between health, environment, and social well-being as vulnerable communities suffered disproportionately from the pandemic. People living in marginalized communities around the world died, faced food and job insecurity, and were excluded from public green spaces at much higher rates than their better-resourced neighbors. Furthermore, citizens demanded that their governments do something about the injustice. In many democratic countries, voters kicked ruling parties out of power and demanded changes: elections held between 2020 and 2022 saw control of government switch parties in Australia, Brazil, Chile, Germany, Italy, South Korea, Sweden, and the United States, to name a few. In non- and quasi-democratic countries, citizens often took to the streets to demand change, as they did in China, Belarus, Iran, Turkey and Russia.

Just as East Asia's developmental states recognized that they would lose legitimacy and political power if they did not accommodate the public's concerns about pollution, it may be that today's eco-developmental states are recognizing that they must add social welfare considerations to their national strategies. It is too early to know for certain, but there are some signs that East

Asia's eco-developmental states may be reframing national policies to prioritize industries and services that offer not just economic and environmental benefits but also social ones.

Japan's vision is perhaps the most developed. Introduced by Prime Minister Abe in 2017, "Society 5.0" offers a vision of a future economic system that is rooted in the UN SDGs rather than maximum economic gain. In this vision, all three bases of sustainability – economy, society, environment – would be equally prioritized. By utilizing advanced technology, societies around the world could include all people in collective prosperity (Holroyd 2022; Narvaez Rojas et al. 2021).[71] Prime Minister Kishida has pledged to increase international aid funding as well as a stimulus package of nearly $500 billion USD, much of it directed toward needy families and green technology (Nohara and Hoirokawa 2022). In Japan, corporate investment in green technology and advanced health care, including AI and robotics, continues to rise (Government of Japan 2021a, 2021b).

While on the campaign trail in 2022, South Korea's new president, Yoon Suk Yeol, also championed a political platform that prioritized environmental sustainability, housing, and new technology that would help the country create a "sympathetic and fair society" (Jones 2022). During his keynote address to the UN General Assembly in September 2022, he argued that "broadening support for the socially disadvantaged groups lays the groundwork for sustainable prosperity."[72] His 2023 budget prioritizes public health and welfare and increases ODA by almost 20 percent (Donor Tracker 2023). Taiwan's President Tsai laid out ambitious economic plans to support industrial growth and green energy transition as well as the largest social housing investment in her 2023 New Year's speech.[73]

Like Japan's Society 5.0 concept, President Xi has also emphasized the need for the development of a new kind of economic system that is more inclusive and sustainable. During his 2022 speech at the G20 meeting in Bali he argued: "We need to build a global partnership for economic recovery, prioritize development and put people at the center" (Xu 2022). As concern for social

---

[71] See also the Japanese Cabinet's Society 5.0 web page, https://www8.cao.go.jp/cstp/english/society5_0/index.html; Keidanren's short video "20XX in Society 5.0: Our Future Created through Digital Transformation," YouTube, April 6, 2020, www.youtube.com/watch?v=cWdGHWfAD1c; and Yuko Harayama's TED Talk about the concept, "Why Society 5.0: Yuko Harayama," YouTube, June 21, 2019, www.youtube.com/watch?v=C2uG2WmMDuA&t=3s.

[72] The text of the speech can be found on the Ministry of Foreign Affairs website: https://overseas.mofa.go.kr/eng/brd/m_5674/view.do?seq=320741#:~:text=Genuine%20peace%20is%20not%20an,want%20of%20energy%20and%20culture.

[73] The official translation of the president's speech can be found on the Office of the President's website: https://english.president.gov.tw/News/6209.

stability in China rises, President Xi has promoted an inclusive green growth model, directing government and private investment into areas that promote social welfare and the environment as well as economic growth (Lin and Zhou 2022; Sun et al. 2020; Zhou, Zhu, and Luo 2022).

It is too soon to know if the regional leaders' visions of an inclusive and sustainable future will be actualized. All four governments' leaders claim to be in support of this vision, and all have put significant financial resources toward these goals, but the future is far from certain. Regional and global security concerns are rising, and China, Japan, South Korea, and Taiwan all increased their defense budgets in 2023. Energy and food insecurity exacerbated by the war in Ukraine may further undermine commitments to an inclusive, green future.

Globally, the dual challenges of the climate crisis and a global pandemic exposed the deep connections between environmental risks, health crises, and social unrest. When the war in Ukraine highlighted the existential security risks of a continued reliance on fossil fuels, extensive efforts to switch to renewable energy and decarbonize economies intensified. Governments around the world, including in East Asia, are funneling trillions of dollars into their economies, prioritizing green technology and a shared and inclusive recovery.

The Intergovernmental Panel on Climate Change (IPCC) has given us fewer than thirty years to make significant changes to the way that we live our lives, build our economies, and run our societies before climate change becomes catastrophic. Change is never easy, but the experience of East Asia has demonstrated that pro-climate policy adjustments are possible. East Asia's experience offers important lessons for the world in how diverse places and people can make rapid, positive headway in improving our local and global environments.

References

Adeola, Francis O. 2000. Cross-National Environmental Injustice and Human Rights Issues: A Review of Evidence in the Developing World. *American Behavioral Scientist* 43 (4): 686–706.

Aivazian, Varouj A., Ying Ge, and Jiaping Qiu. 2005. Can Corporatization Improve the Performance of State-Owned Enterprises Even without Privatization? *Journal of Corporate Finance* 11 (5): 791–808.

Amindoni, Ayomi, and Rebecca Henschke. 2020. The Burning Scar: Inside the Destruction of Asia's Last Rainforests. *BBC News*, November 12.

Amsden, Alice Hoffenberg. 1992. *Asia's Next Giant: South Korea and Late Industrialization*. Oxford: Oxford Academic.

Aspinwall, Nick. 2019. The Tao Indigenous Community Demands Removal of Nuclear Waste from Taiwan's Orchid Island. *The Diplomat*, December 6. https://thediplomat.com/2019/12/tao-indigenous-community-demands-removal-of-nuclear-waste-from-taiwans-orchid-island/.

Avenell, Simon. 2017. *Transnational Japan in the Global Environmental Movement*. Honolulu: University of Hawai'i Press.

——— 2020. Legal Experts and Environmental Activism in Japan. In *Greening East Asia: The Rise of the Eco-developmental State*, edited by Ashley Esarey, Mary Alice Haddad, Joanna Lewis, and Stevan Harrell. Seattle: University of Washington Press, 92–106.

Ayling, Julie, and Neil Gunningham. 2017. Non-state Governance and Climate Policy: The Fossil Fuel Divestment Movement. *Climate Policy* 17 (2): 131–149.

Bamber, Greg J., and Chris J. Leggett. 2001. Changing Employment Relations in the Asia-Pacific Region. *International Journal of Manpower* 22 (4): 300–317.

Baranovitch, Nimrod. 2019. The Impact of Environmental Pollution on Ethnic Unrest in Xinjiang: A Uyghur Perspective. *Modern China* 45 (5): 504–536.

Barbier, Edward B. 2010. Green Stimulus, Green Recovery and Global Imbalances. *World Economics* 11 (2): 149–177.

Bashir, Muhammad Farhan, Muhammad Sadiq, Besma Talbi, Luqman Shahzad, and Muhammad Adnan Bashir. 2022. An Outlook on the Development of Renewable Energy, Policy Measures to Reshape the Current Energy Mix, and How to Achieve Sustainable Economic Growth in the Post COVID-19 Era. *Environmental Science and Pollution Research* 29 (29): 43636–43647.

Berger, Suzanne, and Ronald Dore, eds. 1996. *National Diversity and Global Capitalism*. Ithaca, NY: Cornell University Press.

Biogradlija, Arnes. 2022. ITRI Presents Taiwan Hydrogen Energy Development Roadmap. *H2 Energy News*, July 22.

Bosso, Christopher. 2005. *Environment, Inc.: From Grassroots to Beltway*. Studies in Government and Public Policy. Lawrence: University Press of Kansas.

Bradsher, Keith. 2012. China Asks Other Nations Not to Release Its Air Data. *New York Times*, June 6. www.nytimes.com/2012/06/06/world/asia/china-asks-embassies-to-stop-measuring-air-pollution.html.

Broadbent, Jeffrey. 1998. *Environmental Politics in Japan: Networks of Power and Protest*. New York: Cambridge University Press.

Busch, Per-Olof, Helge Jörgens, and Kerstin Tews. 2005. The Global Diffusion of Regulatory Instruments: The Making of a New International Environmental Regime. *Annals of the American Academy of Political and Social Science* 598 (1): 146–167.

Cabestan, Jean-Pierre. 2017. Changing Identities in Taiwan under Ma Ying-jeou. In *Taiwan and China: Fitful Embrace*, edited by Lowell Dittmer. Oakland: University of California Press, 42–60.

Cai, Wenjia, Can Wang, Jining Chen, and Siqiang Wang. 2011. Green Economy and Green Jobs: Myth or Reality? The Case of China's Power Generation Sector. *Energy* 36 (10): 5994–6003.

Calder, Kent E. 1995. *Strategic Capitalism: Private Business and Public Purpose in Japanese Industrial Finance*. Princeton, NJ: Princeton University Press.

Carr, Anna, Lisa Ruhanen, and Michelle Whitford. 2016. Indigenous Peoples and Tourism: The Challenges and Opportunities for Sustainable Tourism. *Journal of Sustainable Tourism* 24 (8–9): 1067–1079.

Chan, Ta-Chien, Po-Huang Chiang, Ming-Daw Su, Hsuan-Wen Wang, and Michael Shi-yung Liu. 2014. Geographic Disparity in Chronic Obstructive Pulmonary Disease (COPD) Mortality Rates among the Taiwan Population. *PloS One* 9 (5): e98170.

Chang, Hsi-Wen (Lenglengman Rovaniyaw). 2020. Indigenous Attitudes toward Nuclear Waste in Taiwan. In *Greening East Asia: The Rise of the Eco-developmental State*, edited by Ashley Esarey, Mary Alice Haddad, Joanna Lewis, and Stevan Harrell. Seattle: University of Washington Press, 197–212.

Chatzky, Andrew, and James McBride. 2020. *China's Massive Belt and Road Initiative*. Washington, DC: Council on Foreign Relations.

Chen, Wei, Fusheng Wang, Guofeng Xiao, Kai Wu, and Shixuan Zhang. 2015. Air Quality of Beijing and Impacts of the New Ambient Air Quality Standard. *Atmosphere* 6 (8): 1243–1258.

Chen, Xiude, Guocai Chen, Miaoxin Lin, Kai Tang, and Bin Ye. 2022. How Does Anti-corruption Affect Enterprise Green Innovation in China's Energy-Intensive Industries? *Environmental Geochemistry and Health* 44 (9): 2919–2942.

Chen, Yuyu, Avraham Ebenstein, Michael Greenstone, and Hongbin Li. 2013. Evidence on the Impact of Sustained Exposure to Air Pollution on Life Expectancy from China's Huai River Policy. *Proceedings of the National Academy of Sciences* 110 (32): 12936–12941.

Chi, Chun-Chieh. 2001. Capitalist Expansion and Indigenous Land Rights: Emerging Environmental Justice Issues in Taiwan. *The Asia Pacific Journal of Anthropology* 2 (2): 135–153.

Chiu, Yu-Tzu. 2002. Ma Says Taipei's Use of Landfills to End by 2010. *Taipei Times*, July 13.

Cho, Myung-Rae. 2010. The Politics of Urban Nature Restoration: The Case of Cheonggyecheon Restoration in Seoul, Korea. *International Development Planning Review* 32 (2): 145–165.

Choi, Christopher. 2009. Removing Market Barriers to Green Development: Principles and Action Projects to Promote Widespread Adoption of Green Development Practices. *Journal of Sustainable Real Estate* 1 (1): 107–138.

Choo, Hyungsuk, and Tazim Jamal. 2009. Tourism on Organic Farms in South Korea: A New Form of Ecotourism? *Journal of Sustainable Tourism* 17 (4): 431–454.

Chuluu, Khohchahar E. 2021. The Tongpo Case: Indigenous Institutions and Environmental Justice in China. *Critical Asian Studies* 53 (1): 109–125.

Chung, Erin Aeran. 2020. *Immigrant Incorporation in East Asian Democracies*. Cambridge: Cambridge University Press.

Coenen, Johanna, Simon Bager, Patrick Meyfroidt, Jens Newig, and Edward Challies. 2021. Environmental Governance of China's Belt and Road Initiative. *Environmental Policy and Governance* 31 (1): 3–17.

Cui, Shuqin. 2017. Chai Jing's *Under the Dome*: A Multimedia Documentary in the Digital Age. *Journal of Chinese Cinemas* 11 (1): 30–45.

Dal, Emel Parlar, Samiratou Dipama, Saban Çaytas, and Ayda Sezgin. 2021. Assessing the Development–Foreign Policy Nexus of the Asian Rising Powers: South Korea, China, Japan and Indonesia. *Global Policy* 12 (5): 653–662.

Dalton, Russell J. 1994. *The Green Rainbow: Environmental Interest Groups in Western Europe*. New Haven, CT: Yale University Press.

Dasgupta, Susmita, Benoit Laplante, Hua Wang, and David Wheeler. 2002. Confronting the Environmental Kuznets Curve. *The Journal of Economic Perspectives* 16 (1): 147–168.

Deng, Manshu, Wenhong Xie, and Jin Shang. 2022. *China Green Bond Market Report 2021*. Climate Bonds Initiative. www.climatebonds.net/files/reports/cbi_china_sotm_2021_0.pdf.

Ding, Iza. 2020. Performative Governance. *World Politics* 72 (4): 525–556.

Donor Tracker. 2023. South Korea Confirms 18.6% Increase to 2023 ODA Grant Budget. Donor Tracker. https://donortracker.org/policy_updates?policy=south-korea-confirms-18-6-increase-to-2023-oda-grant-budget.

Dore, Ronald Philip. 1973. *British Factory, Japanese Factory: The Origins of National Diversity in Industrial Relations*. Berkeley: University of California Press.

 2000. *Stock Market Capitalism: Welfare Capitalism: Japan and Germany versus the Anglo-Saxons*. Oxford: Oxford Academic.

The Economist. 2014. Business in the Blood: Family Firms. *The Economist (US)* 413 (8911).

Economy, Elizabeth. 2004. *The River Runs Black: The Environmental Challenge to China's Future*. Ithaca, NY: Cornell University Press.

 2011. Roots of Protest and the Party Response. Prepared Statement Before the U.S.-China Economic and Security Review Commission, U.S. Senate/U.S. House of Representatives, First Session, 112th Congress, February 25. www.cfr.org/sites/default/files/pdf/2011/02/Economy.Testimony.2.25.11.pdf.

Eisner, Marc. 2006. *Governing the Environment*. New York: Lynne Rienner.

Ekman, Sigrid-Marianella Stensrud, Huang Wenbin, and Ercilio Langa. 2013. Chinese Trade and Investment in the Mozambican Timber Industry. Center for International Forestry Research Working Paper No. 122. www.cifor.org/publications/pdf_files/WPapers/WP122Ekman.pdf.

Elder, Mark. 2018. Regional Governance for Environmental Sustainability in Asia in the Context of Sustainable Development: A Survey of Regional Cooperation Frameworks. In *Routledge Handbook of Sustainable Development in Asia*, edited by Sara Hsu. New York: Routledge, 468–494.

Elder, Mark, and Peter King, eds. 2018. *Realising the Transformative Potential of the SDGs*. Tokyo: Institute for Global Environmental Strategies.

Epstein, Marc, and Marie-Josée Roy. 1998. Managing Corporate Environmental Performance: A Multinational Perspective. *European Management Journal* 16 (3): 284–296.

Eroğlu, Hasan. 2021. Effects of Covid-19 Outbreak on Environment and Renewable Energy Sector. *Environment, Development and Sustainability* 23 (4): 4782–4790.

Esarey, Ashley, Mary Alice Haddad, Joanna Lewis, and Stevan Harrell, eds. 2020. *Greening East Asia: The Rise of the Eco-developmental State*. Seattle: University of Washington Press.

Esping-Andersen, Gosta. 1990. *Three Worlds of Welfare Capitalism*. Princeton, NJ: Princeton University Press.

Estevez-Abe, Margarita. 2008. *Welfare and Capitalism in Postwar Japan*. New York: Cambridge University Press.

Evans, Peter. 1995. *Embedded Autonomy: States and Industrial Transformation*. Princeton, NJ: Princeton University Press.

Everington, Keoni. 2017. Anti-pollution Protests Held in Taichung and Kaohsiung. *Taiwan News*, December 18. www.taiwannews.com.tw/en/news/3322739.

2020. Chinese Air Pollution to Blow into Taiwan on Friday. *Taiwan News*, October 22. www.taiwannews.com.tw/en/news/4035517.

Fan, Hao, Chuanfeng Zhao, and Yikun Yang. 2020. A Comprehensive Analysis of the Spatio-temporal Variation of Urban Air Pollution in China during 2014–2018. *Atmospheric Environment* 220. https://doi.org/10.1016/j.atmosenv.2019.117066.

Fan, Mei-Fang. 2006. Environmental Justice and Nuclear Waste Conflicts in Taiwan. *Environmental Politics* 15 (3): 417–434.

2021. Indigenous Participation and Knowledge Justice in Deliberative Systems: Flooding and Wild Creek Remediation Controversies in Taiwan. *Environment and Planning C: Politics and Space* 39 (7). https://doi.org/10.1177/23996544211044505.

Fang, Ming, and Weizheng Lai. 2022. Anti-corruption and Political Trust: Evidence from China. https://laiwz.github.io/assets/pdfs/trust_draft.pdf.

Fardoust, Shahrokh, Justin Y. Lin, and Xubei Luo. 2012. Demystifying China's Fiscal Stimulus. World Bank Policy Research Working Paper No. 6221.

Fell, Dafydd. 2017. The Evolution of the Anti-nuclear Movement in Taiwan since 2008. In *Taiwan's Social Movements under Ma Ying-jeou: From the Wild Strawberries to the Sunflowers*, edited by Dafydd Fell. New York: Routledge, 170–192.

2021. *Taiwan's Green Parties: Alternative Politics in Taiwan*. New York: Routledge.

Fisher, Dennis. 2011. Japan Disaster Shakes Up Global Supply Chains. Harvard Business School (website), May 31. https://hbswk.hbs.edu/item/japan-disaster-shakes-up-supply-chain-strategies.

Fraser, Timothy. 2020. Japan's Resilient, Renewable Cities: How Socioeconomics and Local Policy Drive Japan's Renewable Energy Transition. *Environmental Politics* 29 (3): 500–523.

Fuentes-George, Kemi. 2016. *Between Preservation and Exploitation: Transnational Advocacy Networks and Conservation in Developing Countries*. Cambridge, MA: MIT Press.

Fuller, Douglas B. 2008. The Cross-Strait Economic Relationship's Impact on Development in Taiwan and China: Adversaries and Partners. *Asian Survey* 48 (2): 239–264.

Fuyan, Kacaw. 2019. Constructing Ethnic Identity through Taiwanese Indigenous Music and Dance Performances: The Case of TITV. *Te Kaharoa* 13 (3): 1–18.

Gallagher, Mary E. 2002. "Reform and Openness": Why China's Economic Reforms Have Delayed Democracy. *World Politics* 54 (3): 338–372.

Gao, Xiang, and Jessica Teets. 2021. Civil Society Organizations in China: Navigating the Local Government for More Inclusive Environmental Governance. *China Information* 35 (1): 46–66.

Göbel, Christian. 2021. The Political Logic of Protest Repression in China. *Journal of Contemporary China* 30 (128): 169–185.

Goh, Brenda, and Cate Cadell. 2019. China's Xi Says Belt and Road Must Be Green, Sustainable. *Reuters*, April 24.

Goodman, Roger, and Ito Peng. 1996. The East Asian Welfare States: Peripatetic Learning, Adaptive Change, and Nation-Building. In *Welfare States in Transition: National Adaptations in Global Economies*, edited by Gosta Esping-Andersen. London: Sage Publications, 199–224.

Government of Japan. 2021a. Greener, Leaner Growth through Japanese Tech. *Financial Times*, February. www.ft.com/partnercontent/the-government-of-japan/post-covid-innovations-from-japan-3-greener-leaner-growth-through-japanese-tech.html#:~:text=In%20October%202020%2C%20the%20Government,also%20joining%20the%20green%20wave.

2021b. Healthcare Resilience through Japanese Tech. *Financial Times*, February. www.ft.com/partnercontent/the-government-of-japan/post-covid-innovations-from-japan-1-healthcare-resilience-through-japanese-tech.html.

Grano, Simona A. 2015. *Environmental Governance in Taiwan: A New Generation of Activists and Stakeholders*. New York: Routledge.

2020. Interactions between Environmental Civil Society and the State during the Ma Ying-jeou and Tsai Ing-wen Administrations in Taiwan. In *Civil Society and the State in Democratic East Asia: Between Entanglement and Contention in Post High Growth*, edited by David Chiavacci, Simona Grano, and Julia Obinger. Amsterdam: Amsterdam University Press, 34–58.

Guttman, Dan, Oran Young, Yijia Jing et al. 2018. Environmental Governance in China: Interactions between the State and "Nonstate Actors." *Journal of*

Environmental Management 220 (15): 126–135. https://doi.org/10.1016/j.jenvman.2018.04.104.

Haddad, Mary Alice. 2015a. From Backyard Environmental Advocacy to National Democratisation: The Cases of South Korea and Taiwan. In *NIMBY Is Beautiful: Cases of Local Activism and Environmental Innovation around the World*, edited by Carol Hager and Mary Alice Haddad. New York: Berghahn Books, 179–199.

2015b. Increasing Environmental Performance in a Context of Low Governmental Enforcement: Evidence from China. *Journal of Environment and Development* 24 (1): 3–25.

2021. *Effective Advocacy: Lessons from East Asia's Environmentalists.* Cambridge, MA: MIT Press.

Haggard, Stephan. 1990. *Pathways from the Periphery: The Politics of Growth in the Newly Industrializing Countries.* Ithaca, NY: Cornell University Press.

1996. The Political Economy of Regionalism in Asia and the Americas. In *The Political Economy of Regionalism*, edited by Edward D. Mansfield and Helen V. Milner. New York: Columbia University Press, 20–49.

2018. *Developmental States.* New York: Cambridge University Press.

Hall, Derek. 2010. Japanese Lessons and Transnational Forces: ODA and the Environment. In *Japanese Aid and the Construction of the Global Environment: Inescapable Solutions*, edited by David Leheny and Kay Warren. London: Routledge, 167–186.

Hall, Peter, and David Soskice, eds. 2001. *Varieties of Capitalism: The Institutional Foundations of Comparative Advantage.* New York: Oxford University Press.

Han, Heejin. 2015. Korea's Pursuit of Low-Carbon Green Growth: A Middle-Power State's Dream of Becoming a Green Pioneer. *The Pacific Review* 28 (5): 731–754.

Han, Yingwei, and Jie Li. 2022. Should Investors Include Green Bonds in Their Portfolios? Evidence for the USA and Europe. *International Review of Financial Analysis* 80: 101998. https://doi.org/10.1016/j.irfa.2021.101998.

Harris, Paul G., and Graeme Lang, eds. 2015. *Routledge Handbook of Environment and Society in Asia.* New York: Routledge.

Harvey, Fiona. 2021. Richest Nations Agree to End Support for Coal Production Overseas. *The Guardian*, May 21. www.theguardian.com/environment/2021/may/21/richest-nations-agree-to-end-support-for-coal-production-overseas.

Hassard, John, Jonathan Morris, Jackie Sheehan, and Xiao Yuxin. 2010. China's State-Owned Enterprises: Economic Reform and Organizational

Restructuring. *Journal of Organizational Change Management* 23 (5): 500–516.

Hildebrandt, Timothy, and Jennifer Turner. 2009. Green Activism? Reassessing the Role of Environmental NGOs in China. In *State and Society Responses to Social Welfare Needs in China: Serving the People*, edited by Jonathan Schwartz and Shawn Shieh. New York: Routledge, 89–110.

Ho, Ming-sho. 2010. Environmental Movement in Democratizing Taiwan (1980–2004): A Political Opportunity Structure Perspective. In *East Asian Social Movements: Power, Protest, and Change in a Dynamic Region*, edited by J. Broadbent and V. Brockman. New York: Springer, 283–314.

   2014. Resisting Naphtha Crackers: A Historical Survey of Environmental Politics in Taiwan. *China Perspectives* 2014 (3): 5–14.

Holbig, Heike, and Bertram Lang. 2022. China's Overseas NGO Law and the Future of International Civil Society. *Journal of Contemporary Asia* 52 (4): 574–601.

Holroyd, Carin. 2022. Technological Innovation and Building a "Super Smart" Society: Japan's Vision of Society 5.0. *Journal of Asian Public Policy* 15 (1): 18–31.

Hoshi, Takeo, Anil Kashyap, and David Scharfstein. 1991. Corporate Structure, Liquidity, and Investment: Evidence from Japanese Industrial Groups. *The Quarterly Journal of Economics* 106 (1): 33–60.

Hsiao, Hsin-Huang Michael. 1999. Environmental Movements in Taiwan. In *Asia's Environmental Movements: Comparative Perspectives*, edited by Yok-shiu F. Lee and Alvin Y. So. Armonk, NY: M. E. Sharpe, 31–54.

Hsu, Angel, Amy Weinfurter, Jeffrey Tong, and Yihao Xie. 2020. Black and Smelly Waters: How Citizen-Generated Transparency Is Addressing Gaps in China's Environmental Management. *Journal of Environmental Policy & Planning* 22 (1): 138–153.

Hsu, Chia-Hua, and Fang-Yi Cheng. 2019. Synoptic Weather Patterns and Associated Air Pollution in Taiwan. *Aerosol and Air Quality Research* 19 (5): 1139–1151.

Huang, Gillan Chi-Lun, Tim Gray, and Derek Bell. 2013. Environmental Justice of Nuclear Waste Policy in Taiwan: Taipower, Government, and Local Community. *Environment, Development and Sustainability* 15 (6): 1555–1571.

Imura, Hidefumi, and Miranda Schreurs, eds. 2005. *Environmental Policy in Japan*. Northampton, MA: Edward Elgar Publishing.

IPE (Institute of Public & Environmental Affairs) and NRDC (Natural Resources Defense Council). 2014. *Greening the Global Supply Chain:*

CITI Index 2014 Annual Report. Beijing: IPE and NRDC. wwwen.ipe.org
.cn/reports/report_1763.html#.

JETRO (Japan External Trade Organization). 2009. *Japan Views Green Energy
As Essential for Global Economic Recovery*. New York: JETRO. www
.jetro.go.jp/ext_images/en/reports/survey/pdf/2009_04_biz.pdf.

Jia, Zhijie, Shiyan Wen, and Boqiang Lin. 2021. The Effects and Reacts of
COVID-19 Pandemic and International Oil Price on Energy, Economy,
and Environment in China. *Applied Energy* 302: 117612.

Jing, Jun. 2003. Environmental Protests in Rural China. In *Chinese Society:
Change, Conflict, and Resistance*, edited by Elizabeth Perry. New York:
Routledge, 143–160.

Johnson, Chalmers. 1982. *MITI and the Japanese Miracle: The Growth of
Industrial Policy 1925–1975*. Stanford, CA: Stanford University Press.

Johnson, Thomas, Anna Lora-Wainwright, and Jixia Lu. 2018. The Quest for
Environmental Justice in China: Citizen Participation and the Rural–Urban
Network against Panguanying's Waste Incinerator. *Sustainability Science*
13 (3): 733–746.

Jones, Randall. 2022. *President Yoon's Economic Policies*. Seoul: Korea
Economic Institute.

Katz, Richard. 1998. *Japan, The System That Soured: The Rise and Fall of the
Japanese Economic Miracle*. Armonk, NY: M. E. Sharpe.

Kauffmann, Céline, and Camila Saffirio. 2020. Study of International
Regulatory Co-operation (IRC) Arrangements for Air Quality: The
Cases of the Convention on Long-Range Transboundary Air Pollution,
the Canada-United States Air Quality Agreement, and Co-operation in
North East Asia. OECD Regulatory Policy Working Paper No. 12.

Kawai, Masahiro, and Shinji Takagi. 2011. Why Was Japan Hit So Hard by the
Global Financial Crisis? In *The Impact of the Economic Crisis on East
Asia*, edited by Daigee Shaw and Bih Jane Liu. Northampton, MA: Edward
Elgar Publishing, 131–148.

Keck, Margaret, and Kathryn Sikkink. 1998. *Activists beyond Borders:
Advocacy Networks in International Politics*. Ithaca, NY: Cornell
University Press.

Kell, Georg. 2018. The Remarkable Rise of ESG. *Forbes*, July 11.

Kijek, Tomasz. 2015. Modelling of Eco-innovation Diffusion: The EU
Eco-label. *Comparative Economic Research* 18 (1): 65–79.

Kim, Jae Hun, Gunwoo Lee, Ji Young Park, Jungyeol Hong, and
Juneyoung Park. 2019. Consumer Intentions to Purchase Battery Electric
Vehicles in Korea. *Energy Policy* 132: 736–743. https://doi.org/10.1016/
j.enpol.2019.06.028.

Kim, Sung-Young. 2021. National Competitive Advantage and Energy Transitions in Korea and Taiwan. *New Political Economy* 26 (3): 359–375.

Kim, Sung-Young, and Elizabeth Thurbon. 2015. Developmental Environmentalism: Explaining South Korea's ambitious Pursuit of Green Growth. *Politics & Society* 43 (2): 213–240.

King, Gary, Jennifer Pan, and Margaret E. Roberts. 2013. How Censorship in China Allows Government Criticism but Silences Collective Expression. *American Political Science Review* 107 (2): 326–343.

Kraft, Michael, Mark Stephan, and Troy Abel. 2011. *Coming Clean: Information Disclosure and Environmental Performance*. Cambridge, MA: MIT Press.

Ku, Dowan. 1996. The Structural Change of the Korean Environmental Movement. *Korea Journal of Population and Development* 25 (1): 155–180.

2002. Environmental Movement and Policies During High Economic Growth in Korea. In *Environment and Our Sustainability in the 21st Century: Understanding and Cooperation between Developed and Developing Countries*, edited by Yuko Arayama. Nagoya: Nagoya University, 65–87.

2011. The Korean Environmental Movement: Green Politics through Social Movement. In *East Asian Social Movements*, edited by Jeffrey Broadbent and Vicky Brockman. New York: Springer, 205–235.

Ladislaw, Sarah O., and Nitzan Goldberger. 2010. Assessing the Global Green Stimulus. Center for Strategic and International Studies, February 2016. https://csis-website-prod.s3.amazonaws.com/s3fs-public/legacy_files/files/publication/010216_Ladislaw_GlobalGreenStimulus_0.pdf.

Lardy, Nicholas R., and Arvind Subramanian. 2011. *Sustaining China's Economic Growth after the Global Financial Crisis*. Peterson Institute.

Lawrence, Martha, Richard Bullock, and Ziming Liu. 2019. *China's High-Speed Rail Development*. Washington, DC: World Bank Publications.

Lee, Jae-Hyup, and Jisuk Woo. 2020. Green New Deal Policy of South Korea: Policy Innovation for a Sustainability Transition. *Sustainability* 12 (23): 10191.

Lee, Jong Youl, and Chad David Anderson. 2013. The Restored Cheonggyecheon and the Quality of Life in Seoul. *Journal of Urban Technology* 20 (4): 3–22.

Lee, See-Jae. 2000. The Environmental Movement and Its Political Empowerment. *Korea Journal* 40 (3): 131–160.

Lee, Su-Hoon, Hsin-Huang Michael Hsiao, Hwa-Jen Liu et al. 1999. The Impact of Democratization on Environmental Movements. In *Asia's*

Environmental Movements, edited by Yok-shiu F. Lee and Alvin Y. So. Armonk, NY: M. E. Sharpe, 230–251.

Lee, Young-Sook, Laura J. Lawton, and David B. Weaver. 2013. Evidence for a South Korean Model of Ecotourism. *Journal of Travel Research* 52 (4): 520–533.

Lehney, David. 2003. *The Rules of Play: National Identity and the Shaping of Japanese Leisure*. Ithaca, NY: Cornell University Press.

Lengauer, Sara. 2011. China's Foreign Aid Policy: Motive and Method. *Culture Mandala* 9 (2): 5899.

Leonard, Pauline, and Angela Lehmann. 2019. International Migrants in China: Civility, Contradiction, and Confusion. In *Destination China*, edited by Angela Lehmann and Pauline Leonard. New York: Springer, 1–17.

Lewallen, Ann-elise. 2016. Signifying Ainu Space: Reimagining Shiretoko's Landscapes through Indigenous Ecotourism. *Humanities* 5 (3): 1–11. https://doi.org/10.3390/h5030059.

Li, Junxia. 2019. China ISO14001 Certification Overview and Literature Review. *Journal of Human Resource and Sustainability Studies* 7 (1): 108–120.

Lin, Boqiang, and Yicheng Zhou. 2022. Measuring the Green Economic Growth in China: Influencing Factors and Policy Perspectives. *Energy* 241: 122518.

Lin, Chin-Huang, Ho-Li Yang, and Dian-Yan Liou. 2009. The Impact of Corporate Social Responsibility on Financial Performance: Evidence from Business in Taiwan. *Technology in Society* 31 (1): 56–63.

Lin, Songhua, and Alyson C. Ma. 2012. Outsourcing and Productivity: Evidence from Korean Data. *Journal of Asian Economics* 23 (1): 39–49.

Lindenberg, Nannette. 2014. Definition of Green Finance. German Development Institute proposal. www.cbd.int/financial/gcf/definition-greenfinance.pdf.

Liu, Dongshu. 2020. Advocacy Channels and Political Resource Dependence in Authoritarianism: Nongovernmental Organizations and Environmental Policies in China. *Governance* 33 (2): 323–342.

Liu, Qianqian. 2016. Japan's Practice in Green Aid and Its Implications for China. *China Quarterly of International Strategic Studies* 2 (2): 201–217.

Liu, Shuang, Ye Wang, and Yan Wang. 2021. South Korea and Japan Will End Overseas Coal Financing. Will China Catch Up? World Resources Institute, June 14. www.wri.org/insights/south-korea-and-japan-will-end-overseas-coal-financing-will-china-catch.

Liu, Zuankuo, and Li Xin. 2019. Has China's Belt and Road Initiative Promoted Its Green Total Factor Productivity? Evidence from Primary Provinces along the Route. *Energy Policy* 129: 360–369.

Luo, Chen, Anfan Chen, Botao Cui, and Wang Liao. 2021. Exploring Public Perceptions of the COVID-19 Vaccine Online from a Cultural Perspective: Semantic Network Analysis of Two Social Media Platforms in the United States and China. *Telematics and Informatics* 65: 101712. https://doi.org/10.1016/j.tele.2021.101712.

Ma, Tianjie. 2008. Environmental Mass Incidents in Rural China: Examining Large-Scale Unrest in Dongyang, Zhejiang. *China Environment Series*: 33–56. www.wilsoncenter.org/sites/default/files/media/documents/publication/ma_feature_ces10.pdf.

Makki, Anas A., Hisham Alidrisi, Asif Iqbal, and Basil O Al-Sasi. 2020. Barriers to Green Entrepreneurship: An ISM-Based Investigation. *Journal of Risk and Financial Management* 13 (11): 249–266.

Manion, Melanie. 2000. Chinese Democratization in Perspective: Electorates and Selectorates at the Township Level. *The China Quarterly* 163: 764–782.

Maruyama, Hiroshi. 2012. Ainu Landowners' Struggle for Justice and the Illegitimacy of the Nibutani Dam Project in Hokkaido Japan. *International Community Law Review* 14 (1): 63–80.

McKean, Margaret. 1981. *Environmental Protest and Citizen Politics in Japan.* Berkeley: University of California.

McKerracher, Colin. 2022. China Has Shot at Seizing 60% Share of Global EV Sales This Year. *Bloomberg*, November 15.

Mertha, Andrew. 2008. *China's Water Warriors: Citizen Action and Policy Change.* Ithaca, NY: Cornell University Press.

Mufson, Steven. 2021. Huge Plastics Plant Faces Calls for Environmental Justice, Stiff Economic Headwinds. *The Washington Post*, April 19.

Murdie, Amanda, and Johannes Urpelainen. 2015. Why Pick On Us? Environmental INGOs and State Shaming As a Strategic Substitute. *Political Studies* 63 (2): 353–372.

Murphy, Matt. 2022. China's Protests: Blank Paper Becomes the Symbol of Rare Demonstrations. *BBC News*, November 28.

Nakamura, Eri. 2011. Does Environmental Investment Really Contribute to Firm Performance? An Empirical Analysis Using Japanese Firms. *Eurasian Business Review* 1 (2): 91–111.

Narvaez Rojas, Carolina, Gustavo Adolfo Alomia Peñafiel, Diego Fernando Loaiza Buitrago, and Carlos Andrés Tavera Romero. 2021. Society 5.0: A Japanese Concept for a Superintelligent Society. *Sustainability* 13 (12): 1–17.

Naughton, Barry. 2009. Understanding the Chinese Stimulus Package. *China Leadership Monitor* 28 (2): 1–12.

2017. The General Secretary's Extended Reach: Xi Jinping Combines Economics and Politics. Hoover Institution, September 11. www.hoover .org/research/general-secretarys-extended-reach-xi-jinping-combines-eco nomics-and-politics.

Nedopil, Christoph. 2021. *Investments in the Chinese Belt and Road Initiative (BRI) in 2020: A Year of COVID-19*. Beijing: International Institute of Green Finance. https://greenfdc.org/wp-content/uploads/2021/01/China-BRI-Investment-Report-2020.pdf.

Nie, Changfei, Yajing Zhou, and Yuan Feng. 2022. Can Anti-corruption Induce Green Technology Innovation? Evidence from a Quasi-natural Experiment of China. *Environmental Science and Pollution Research* 30 (12): 34932–34951.

Noble, Gregory W., and John Ravenhill. 2000. *The Asian Financial Crisis and the Architecture of Global Finance*. Cambridge: Cambridge University Press.

Nohara, Yoshiaki, and Takashi Hoirokawa. 2022. Japan's Kishida Spends Big to East Inflation, Popularity Woes. *Bloomberg*, October 27.

OECD (Organisation for Economic Co-operation and Development). 2009. Policy Responses to the Economic Crisis: Investing in Innovation for Long-Term Growth. Paris: OECD.

2016. *Green Growth in Hai Phong, Viet Nam*. Paris: OECD.

Oh, Jae In, Hyungkyoo Kim, and Dongwook Sohn. 2020. Minority Neighbourhoods and Availability of Green Amenities: Empirical Findings from Seoul, South Korea. *Local Environment* 25 (1): 69–82.

Okano-Heijmans, Maaike. 2012. Japan's "Green" Economic Diplomacy: Environmental and Energy Technology and Foreign Relations. *The Pacific Review* 25 (3): 339–364.

O'Neill, Michael. 1997. *Green Parties and Political Change in Contemporary Europe: New Politics, Old Predicaments*. Burlington, VT: Ashgate.

Öniş, Ziya. 1991. The Logic of the Developmental State. *Comparative Politics* 24 (1): 109–126.

Ostrom, Elinor. 1990. *Governing the Commons: The Evolution of Institutions for Collective Action*. New York: Cambridge University Press.

Pearson, Margaret, Meg Rithmire, and Kellee S. Tsai. 2021. Party-State Capitalism in China. *Current History* 120 (827): 207–213.

2023. *The State and Capitalism in China*. Elements in Politics and Society in East Asia. New York: Cambridge University Press.

Pirie, Iain. 2018. Korea and Taiwan: The Crisis of Investment-Led Growth and the End of the Developmental State. *Journal of Contemporary Asia* 48 (1): 133–158.

Pitrelli, Monica. 2022. Asia-Pacific's Travel Industry Could Be the First to Recover by 2023. *CNBC*, October 17.

Porter, Michael E. 1990. *The Competitive Advantage of Nations*. New York: Free Press.

Poulos, Helen, and Mary Alice Haddad. 2016. Violent Repression of Environmental Protests. *SpringerPlus* 5 (230): 1–12.

Prakash, Aseem, and Matthew Potoski. 2007. Investing Up: FDI and the Cross-Country Diffusion of ISO 14001 Management Systems. *International Studies Quarterly* 51 (3): 723–744.

Qiu, Lu, Xiaowen Jie, Yanan Wang, and Minjuan Zhao. 2020. Green Product Innovation, Green Dynamic Capability, and Competitive Advantage: Evidence from Chinese Manufacturing Enterprises. *Corporate Social Responsibility and Environmental Management* 27 (1): 146–165.

Rawski, Thomas. 1999. Reforming China's Economy: What Have We Learned? *The China Journal* 41: 139–156.

Reardon-Anderson, James. 1997. *Pollution, Politics, and Foreign Investment in Taiwan: The Lukang Rebellion*. Armonk, NY: M. E. Sharpe.

Reimann, Kim. 1999. Building Networks from the Outside In: International Movements, Japanese NGOs and the Kyoto Climate Change Conference. Paper Presented at the Annual Meeting of the Northeastern Political Science Association and International Studies Association, Philadelphia, PA, November 10–14.

2003. Building Global Civil Society from the Outside In? Japanese International Development NGOs, the State, and International Norms. In *The State of Civil Society in Japan*, edited by Frank Schwartz and Susan Pharr. New York: Cambridge University Press, 298–315.

Republic of Korea. 2011. *Low Carbon, Green Growth: Korea's Third National Communication under the United Nations Framework Convention on Climate Change*. Seoul: United Nations Framework Convention on Climate Change.

Roberts, Stephen L., and Ilan Kelman. 2022. Global Health Security and Islands As Seen through COVID-19 and Vaccination. *Global Public Health* 17 (4): 601–613.

Robinson, Mary. 2018. *Climate Justice: Hope, Resilience, and the Fight for a Sustainable Future*. New York: Bloomsbury Publishing.

Rodó, Xavier, Roger Curcoll, Marguerite Robinson et al. 2014. Tropospheric Winds from Northeastern China Carry the Etiologic Agent of Kawasaki Disease from Its Source to Japan. *Proceedings of the National Academy of Sciences* 111 (22): 7952–7957.

Rodrigues, Maria Guadalupe Moog. 2003. *Global Environmentalism and Local Politics: Transnational Advocacy Networks in Brazil, Ecuador, and India*. Albany: State University of New York Press.

Rozenas, Arturas. 2020. A Theory of Demographically Targeted Repression. *Journal of Conflict Resolution* 64 (7–8): 1254–1278.

Sanchez, Diego. 2020. Amazon Community Files Lawsuit against Chinese Firm over Gas Flaring. Phys.org, December 11. https://phys.org/news/2020-12-amazon-lawsuit-chinese-firm-gas.html.

Schaede, Ulrike, and Kay Shimizu. 2022. *The Digital Transformation and Japan's Political Economy*. Elements in Politics and Society in East Asia. New York: Cambridge University Press.

Schlosberg, David. 2009. *Defining Environmental Justice: Theories, Movements, and Nature*. Oxford: Oxford University Press.

Schreurs, Miranda. 2002. *Environmental Politics in Japan, Germany, and the United States*. New York: Cambridge University Press.

2005. Japan and Global Environmental Governance. In *Contested Governance in Japan: Sites and Issues*, edited by Glenn Hook. New York: Routledge, 157–175.

Schwartz, Frank J. 1998. *Advice and Consent: The Politics of Consultation in Japan*. New York: Cambridge University Press.

Scull, Erika. 2008. Environmental Health Challenges in Xinjiang. In *Research Brief Produced As Part of the China Environment Forums Partnership with Western Kentucky University on the USAID-Supported China Environmental Health Project, Western Kentucky*. Washington DC: Wilson Center.

Shapiro, Judith. 2019. China's Environmental Challenges. In *Green Planet Blues: Critical Perspectives on Global Environmental Politics*, edited by Ken Conca and Geoffrey Dabelko. New York: Routledge, 101–108.

Shie, Yi-Jen. 2020. Indigenous Legacy for Building Resilience: A Case Study of Taiwanese Mountain River Ecotourism. *Tourism Management Perspectives* 33: 100612. https://doi.org/10.1016/j.tmp.2019.100612.

Shieh, Shawn. 2018. The Chinese State and Overseas NGOs: From Regulatory Ambiguity to the Overseas NGO Law. Paper Presented at the Nonprofit Policy Forum.

Shinkawa, Toshimitsu, and T. J. Pempel. 1996. Occupational Welfare and the Japanese Experience. In *The Privatization of Social Policy? Occupational Welfare and the Welfare State in America, Scandinavia and Japan*, edited by Michael Shalev. New York: Routledge, 280–326.

Siciliano, Giuseppina, Daniela Del Bene, Arnim Scheidel, Juan Liu, and Frauke Urban. 2019. Environmental Justice and Chinese Dam-Building in the Global South. *Current Opinion in Environmental Sustainability* 37: 20–27.

Sidel, Mark. 2019. Managing the Foreign: The Drive to Securitize Foreign Nonprofit and Foundation Management in China. *VOLUNTAS: International Journal of Voluntary and Nonprofit Organizations* 30 (4): 664–677.

Simeon, Roblyn. 2010. Evaluating the Strategic Implications of Japanese IT Offshore Outsourcing in China and India. *International Journal of Management & Information Systems (IJMIS)* 14 (3): 25–36.

Singleton, Deborah, and Angela Su. 2016. How China's 13th Five-Year Plan Addresses Energy and the Environment. *ChinaFile*, March 10. www.chinafile.com/reporting-opinion/environment/how-chinas-13th-five-year-plan-addresses-energy-and-environment.

Siniawer, Eiko Maruko. 2018. A War against Garbage in Postwar Japan. *Asia-Pacific Journal: Japan Focus* 16 (22): 1–12.

Snape, Holly. 2021. Cultivate Aridity and Deprive Them of Air: Altering the Approach to Non-State-Approved Social Organizations. *Made in China Journal* 6 (1): 54–59.

Sommer, Martin. 2009. Why Has Japan Been Hit So Hard by the Global Recession? *IMF Staff Position Notes* 2009 (5). www.imf.org/external/pubs/ft/spn/2009/spn0905.pdf.

Spires, Anthony J. 2020. Regulation As Political Control: China's First Charity Law and Its Implications for Civil Society. *Nonprofit and Voluntary Sector Quarterly* 49 (3): 571–588.

Steger, Isabella. 2020. China Tried to Threaten Taiwan by Weaponizing Tourism, but It Didn't Work. *Quartz*, January 7.

Sun, Yuhuan, Wangwang Ding, Zhiyu Yang, Guangchun Yang, and Juntao Du. 2020. Measuring China's Regional Inclusive Green Growth. *Science of the Total Environment* 713: 136367.

Sun, Yunpeng, Weimin Guan, Yuning Cao, and Qun Bao. 2022. Role of Green Finance Policy in Renewable Energy Deployment for Carbon Neutrality: Evidence from China. *Renewable Energy* 197 ©: 643–653. https://econpapers.repec.org/article/eeerenene/v_3a197_3ay_3a2022_3ai_3ac_3ap_3a643-653.htm.

Tang, Shui-Yan, and Ching-Ping Tang. 1997. Democratization and Environmental Politics in Taiwan. *Asian Survey* 37 (3): 281–294.

Tian, Jinfang, Longguang Yu, Rui Xue, Shan Zhuang, and Yuli Shan. 2022. Global Low-Carbon Energy Transition in the Post-COVID-19 Era. *Applied Energy* 307: 118205.

Tiberghien, Yves. 2021. *The East Asian Covid-19 Paradox*. Elements in Politics and Society in East Asia. Cambridge: Cambridge University Press.

Ting-jieh, Wang. 2021. Indigenous Peoples and the Politics of the Environment in Taiwan. In *Taiwan's Contemporary Indigenous Peoples*, edited by Chia-yuan Huang and Dafydd Fell. Abingdon: Routledge, 223–38.

Tokunaga, Shojiro. 1992. Japan's FDI-Promoting Systems and Intra-Asian Networks: New Investment and Trade Systems Created by the Borderless Economy. In *Japan's Foreign Investment and Asian Economic Interdependence Production, Trade, and Financial Systems*, edited by Shojiro Tokunaga. Tokyo: University of Tokyo Press, 5–47.

Tu, Wen-Ling. 2019. Combating Air Pollution through Data Generation and Reinterpretation: Community Air Monitoring in Taiwan. *East Asian Science, Technology and Society: An International Journal* 13 (2): 235–255.

Turner, Jennifer, and Linden Ellis. 2007. China's Green Olympics: A Lasting Impact? Wilson Center. www.wilsoncenter.org/event/chinas-green-olym pics-lasting-impact#:~:text=Being's%20green%20Olympic%20efforts% 20have,illnesses%20caused%20by%20air%20pollution.

UN Inter-agency Task Force on Financing for Development. 2020. *Financing for Sustainable Development Report 2020*. New York: United Nations.

UNEP (United Nations Environmental Program). 2009. Independent Environmental Assessment, Beijing 2008 Olympic Games. New York: United Nations.

Upham, Frank. 1976. Litigation and Moral Consciousness in Japan: An Interpretive Analysis of Four Japanese Pollution Suits. *Law and Society Review* 10 (4): 579–619.

　1987. *Law and Social Change in Postwar Japan* Cambridge, MA: Harvard University Press.

Uriu, Robert M. 2021. Betting on Hydrogen: Japan's Green Industrial Policy for Hydrogen and Fuel Cells. In *New Challenges and Solutions for Renewable Energy: Japan, East Asia and Northern Europe*, edited by Paul Midford and Espen Moe. New York: Palgrave Macmillan, 149–180.

Van Rooij, Benjamin. 2010. The People vs. Pollution: Understanding Citizen Action against Pollution in China. *Journal of Contemporary China* 19 (63): 55–77.

Vogel, Steven K. 1996. *Freer Markets, More Rules: Regulatory Reform in Advanced Industrial Countries*. Ithaca, NY: Cornell University Press.

Volodzko, David Josef. 2017. China Is to Blame for Korea's Pollution? Really? *South China Morning Post*, April 15. www.scmp.com/week-asia/politics/ article/2087447/china-blame-koreas-pollution-really.

Wade, Robert. 1990. *Governing the Market: Economic Theory and the Role of Government in East Asian Industrialization*. Princeton, NJ: Princeton University Press.

Walker, Brett. 2011. *Toxic Archipelago: A History of Industrial Disease in Japan*. Weyerhaeuser Environmental Books. Seattle: University of Washington Press.

Wang, Alex L. 2013. The Search for Sustainable Legitimacy: Environmental Law and Bureaucracy in China. *Harvard Environmental Law Re*view 37 (2): 365–440.

Watanabe, Takehiro. 2013. Talking Sulfur Dioxide: Air Pollution and the Politics of Science in Late Meiji Japan. In *Japan at Nature's Edge: The Environmental Context of a Global Power*, edited by Ian J. Miller, Julia A. Thomas and Brett L. Walker. Honolulu: University of Hawai'i Press, 73–89.

Williams, Michelle. 2014. *The End of the Developmental State?* New York: Routledge.

Wolch, Jennifer R., Jason Byrne, and Joshua P. Newell. 2014. Urban Green Space, Public Health, and Environmental Justice: The Challenge of Making Cities "Just Green Enough." *Landscape and Urban Planning* 125: 234–244.

Woodall, Brian. 1996. *Japan under Construction: Corruption, Politics, and Public Works*. Berkeley: University of California Press.

World Bank. 2021. *Case Study on Korea's Green Growth Recovery*. Seoul: World Bank.

Xia, Cai. 2022. The Weakness of Xi Jinping: How Hubris and Paranoia Threaten China's Future. *Foreign Affairs* 101: 85.

Xiao, Qiang. 2019. The Road to Digital Unfreedom: President Xi's Surveillance State. *Journal of Democracy* 30 (1): 53–67.

Xie, Dennis. 2020. Groups Announce Air Pollution Protest. *Taipei Times*, July 10. www.taipeitimes.com/News/taiwan/archives/2020/07/10/2003739689.

Xie, Zhenhua. 2020. China's Historical Evolution of Environmental Protection along with the Forty Years' Reform and Opening-Up. *Environmental Science and Ecotechnology* 1. www.sciencedirect.com/science/article/pii/S2666498419300018/pdfft?md5=197139aec74e41d2c047f86892756256&pid=1-s2.0-S2666498419300018-main.pdf.

Xin, Ming. 2018. *China's New Strategic Layout*. Singapore: Springer.

Xing, Yu-Fei, Yue-Hua Xu, Min-Hua Shi, and Yi-Xin Lian. 2016. The Impact of $PM_{2.5}$ on the Human Respiratory System. *Journal of Thoracic Disease* 8 (1): E69.

Xing, Yunfei, Yuhai Li, and Feng-Kwei Wang. 2021. How Privacy Concerns and Cultural Differences Affect Public Opinion during the COVID-19 Pandemic: A Case Study. *Aslib Journal of Information Management* 73 (4): 517–542.

Xinhua. 2020. Chinese Tourists Make Over 6 bln Domestic Trips in 2019. *China Daily*, March 11.

Xu, Wei. 2022. Xi Stresses Need for Inclusive Growth. *China Daily*, November 16.

Yang, Guobin. 2005. Environmental NGOs and Institutional Dynamics in China. *The China Quarterly* 181: 46–66.

Yang, Xueyan. 2019. Participatory Management of Community-Based Ecotourism at Jiuzhaigou National Nature Reserve, China. Ph.D. dissertation, Brandenburg University of Technology.

Yarime, Masaru, and Martin Karlsson. 2018. Understanding the Innovation System of Smart Cities: The Case of Japan and Implications for Public Policy and Institutional Design. In *Innovation Policy, Systems and Management*, edited by Jorge Niosi. New York: Cambridge University Press, 394–417.

Yasumoto, Shinya, Tomoki Nakaya, and Andrew P Jones. 2020. Quantitative Environmental Equity Analysis of Perceived Accessibility to Urban Parks in Osaka Prefecture, Japan. *Applied Spatial Analysis and Policy* 14: 337–354.

Yeh, Yin-hua, Tsun-siou Lee, and Tracie Woidtke. 2001. Family Control and Corporate Governance: Evidence from Taiwan. *International Review of Finance* 2 (1–2): 21–48.

Yim, Steve Hung Lam, Yefu Gu, Matthew A. Shapiro, and Brent Stephens. 2019. Air Quality and Acid Deposition Impacts of Local Emissions and Transboundary Air Pollution in Japan and South Korea. *Atmospheric Chemistry and Physics* 19 (20): 13309–13323.

Yokohama, Julian Ryall. 2013. Hazy Days in Japan. *Deutsche Welle*, December 3. www.dw.com/en/parts-of-japan-smothered-in-chinese-air-pollution/a-16665471.

Yoon, D. K., Jung Eun Kang, and Juhyeon Park. 2017. Exploring Environmental Inequity in South Korea: An Analysis of the Distribution of Toxic Release Inventory (TRI) Facilities and Toxic Releases. *Sustainability* 9 (10): 1886. https://doi.org/10.3390/su9101886.

Zameer, Hashim, Ying Wang, and Humaira Yasmeen. 2020. Reinforcing Green Competitive Advantage through Green Production, Creativity and Green Brand Image: Implications for Cleaner Production in China. *Journal of Cleaner Production* 247: 119119.

Zelenovskaya, Ekaterina. 2012. *Green Growth Policy in Korea: A Case Study.* Venice: International Center for Climate Governance. www.greenpolicyplat

form.org/sites/default/files/downloads/resource//GreenGrowthPolicyKorea
.ICCG_.pdf.

Zeppel, Heather. 2007. Indigenous Ecotourism: Conservation and Resource Rights. In *Critical Issues in Ecotourism*, edited by James Higham. Oxford: Butterworth-Heinemann, 308–348.

Zhang, Dongyang, Muhammad Mohsin, Abdul Khaliq Rasheed, Youngho Chang, and Farhad Taghizadeh-Hesary. 2021. Public Spending and Green Economic Growth in BRI Region: Mediating Role of Green Finance. *Energy Policy* 153: 112256.

Zhang, Falin. 2018. The Chinese Developmental State: Standard Accounts and New Characteristics. *Journal of International Relations and Development* 21: 739–768.

Zhang, Zhenguo, and Lee Liu. 2021. Environmental Health and Justice in a Chinese Environmental Model City. *Journal of Environmental Health* 83 (6): 30–38.

Zheng, Yu. 2022. The Third Way of Inclusive Growth in China. *Asian Review of Political Economy* 1 (3): 1–16.

Zhou, Guangyou, Jieyu Zhu, and Sumei Luo. 2022. The Impact of Fintech Innovation on Green Growth in China: Mediating Effect of Green Finance. *Ecological Economics* 193: 107308.

Zhou, Lihuan, Sean Gilbert, Ye Wang, Miquel Muñoz Cabré, and Kevin P Gallagher. 2018. Moving the Green Belt and Road Initiative: From Words to Actions. Washington, DC: World Resources Institute. www.wri.org/research/moving-green-belt-and-road-initiative-words-actions.

Zhuang, Lixia, Tracy Taylor, David Beirman, and Simon Darcy. 2017. Socially Sustainable Ethnic Tourism: A Comparative Study of two Hakka Communities in China. *Tourism Recreation Research* 42 (4): 467–483.

Zissis, Carin, and Jayshree Bajoria. 2008. *China's Environmental Crisis*. Washington, DC: Council on Foreign Relations.

Cambridge Elements ⹀

Politics and Society in East Asia

Erin Aeran Chung

The Johns Hopkins University

Erin Aeran Chung is the Charles D. Miller Professor of East Asian Politics in the Department of Political Science at the Johns Hopkins University. She specializes in East Asian political economy, migration and citizenship, and comparative racial politics. She is the author of *Immigration and Citizenship in Japan* (Cambridge, 2010, 2014; Japanese translation, Akashi Shoten, 2012) and *Immigrant Incorporation in East Asian Democracies* (Cambridge, 2020). Her research has been supported by grants from the Academy of Korean Studies, the Japan Foundation, the Japan Foundation Center for Global Partnership, the Social Science Research Council, and the American Council of Learned Societies.

Mary Alice Haddad

Wesleyan University

Mary Alice Haddad is the John E. Andrus Professor of Government, East Asian Studies, and Environmental Studies at Wesleyan University. Her research focuses on democracy, civil society, and environmental politics in East Asia as well as city diplomacy around the globe. A Fulbright and Harvard Academy scholar, Haddad is author of *Effective Advocacy: Lessons from East Asia's Environmentalists* (MIT, 2021), *Building Democracy in Japan* (Cambridge, 2012), and *Politics and Volunteering in Japan* (Cambridge, 2007), and co-editor of *Greening East Asia* (University of Washington, 2021), and *NIMBY is Beautiful* (Berghahn Books, 2015). She has published in journals such as *Comparative Political Studies, Democratization, Journal of Asian Studies,* and *Nonprofit and Voluntary Sector Quarterly,* with writing for the public appearing in the *Asahi Shimbun,* the *Hartford Courant,* and the *South China Morning Post.*

Benjamin L. Read

University of California, Santa Cruz

Benjamin L. Read is a professor of Politics at the University of California, Santa Cruz. His research has focused on local politics in China and Taiwan, and he also writes about Issues and techniques in comparison and field research. He is author of *Roots of the State: Neighborhood Organization and Social Networks in Beijing and Taipei* (Stanford, 2012), coauthor of *Field Research in Political Science: Practices and Principles* (Cambridge, 2015), and co-editor of *Local Organizations and Urban Governance in East and Southeast Asia: Straddling State and Society* (Routledge, 2009). His work has appeared in journals such as *Comparative Political Studies, Comparative Politics, the Journal of Conflict Resolution, the China Journal, the China Quarterly,* and *the Washington Quarterly,* as well as several edited books.

About the Series

The Cambridge Elements series on Politics and Society in East Asia offers original, multidisciplinary contributions on enduring and emerging issues in the dynamic region of East Asia by leading scholars in the field. Suitable for general readers and specialists alike, these short, peer-reviewed volumes examine common challenges and patterns within the region while identifying key differences between countries. The series consists of two types of contributions: 1) authoritative field surveys of established concepts and themes that offer roadmaps for further research; and 2) new research on emerging issues that challenge conventional understandings of East Asian politics and society. Whether focusing on an individual country or spanning the region, the contributions in this series connect regional trends with points of theoretical debate in the social sciences and will stimulate productive interchanges among students, researchers, and practitioners alike.

.

Cambridge Elements $\equiv$

Politics and Society in East Asia